SOTHIS PRESS

Sothis Press aims to completely reimagine the public domain book with an emphasis on universal access to captivatingly designed print editions. We believe that technology will enable every single book ever published to be perpetually accessible in beautiful print form. Sothis is beauty in ubiquity.

www.sothis.press
1 Waterhouse Square, London, ECIN 2ST

The Construction of Mill Dams

Comprising Also the Building of Race and Reservoir Embankments and Head Gates, the Measurement of Streams, Gauging of Water Supply, &C

James Leffel

SOTHIS PRESS

ISBN 978-1-0368-0906-5

This edition is published by
Sothis Press © 2023. The company
is registered in England and Wales.
The Company Number is 14900389.

Clarence Johnston.

63 S. 4th ave.

Ann Arbor, Mich

CONSTRUCTION OF

MILL DAMS:

COMPRISING ALSO

THE BUILDING OF RACE AND RESERVOIR EMBANKMENTS AND HEAD GATES, THE MEASUREMENT OF STREAMS, GAUGING OF WATER SUPPLY, &C.

Illustrated by numerous Full-Page Plates.

SPRINGFIELD, OHIO:
JAMES LEFFEL & CO.,
AUTHORS AND PUBLISHERS.
1874.

1

PREFACE.

A considerable portion of the matter contained in this work has appeared in a serial form in the monthly issues of Leffel's Illustrated Milling and Mechanical News, but has been carefully revised and corrected for republication in this form; while the remaining chapters and illustrations are entirely new, being now printed for the first time. While the series of articles on this subject was in progress in the monthly journal referred to, the degree of interest with which it was received by the milling public was indicated by frequent requests from all parts of the country for files of the paper containing the complete series. The number and the urgent character of these applications, with most of which the publishers were unable to comply, the earlier editions of the paper being soon exhausted, suggested the expediency of publishing the series entire, in a form which would admit of fully supplying both the present and future demand. It may also be said, in further justification of this course, that in view of all the circumstances of the case and the amount of labor and outlay required in the preparation of these articles with the illustrations accompanying them, it seemed desirable that the results attained should be put in more permanent shape than it is possible to give them in a periodical publication. The collection of materials for this work occupied a period of over three years, and the facts embodied in it were gathered from all sections of our country, so that for American readers at least we may venture to hope that it will not be found wanting in comprehensiveness and variety.

Abstract and technical features, it will be observed, have been uniformly avoided, our object being to make this volume useful in the highest possible degree to every person owning water power or interested in any way in its utilization. To this end we have chosen the presentation of the subjects discussed in this volume in the most direct and practical form; and whatever may be its deficiencies in other respects, we trust that by minuteness of detail and distinctness of description it may be rendered serviceable to the class of readers to whose favor it is respectfully commended.

JAMES LEFFEL & CO.

SPRINGFIELD, OHIO.

CONTENTS.

INTRODUCTION.

The household proverb that "Fire is a good servant, but a bad master," is true in an almost equal degree of water, its opposing element. Under the intelligent control of man, the power of water is one of the chief instrumentalities in promoting human comfort and spreading the blessings of civilization. The uses to which it may be turned are so various, and the products for the manufacture of which it is a serviceable agent are of such manifold forms and minister to so many wants of our nature, that the country which is possessed of abundant water power is looked upon as especially favored by Providence. But this invaluable servant, if once it bursts over the bounds which have been set for its action, becomes a destructive scourge, laying waste the very fields whose tillage it had made profitable and bringing to poverty, in an hour as it were, a whole community, whom until then it had sustained in prosperous industry. More than one such instance occurs in our own memory and doubtless in that of every reader, where an insignificant stream along whose banks thousands of men, women and children have earned for years a secure livelihood, has risen in its might and swept away with its swollen torrent all the fruits of those years of labor, often not sparing even the life of the laborer himself, and in a single day inflicting damage which a generation may not see repaired.

As the first step in the improvement of a water power is the construction of the dam, so the first consideration in that work is strength and durability. While the builder of a mill is bound to consult economy in the means he employs in obtaining a sufficient fall for the purposes of his business, and diverting from its natural channel that portion of the stream which is to be made directly available, the economy which does not provide a sure resistance to sudden floods is very apt to prove the worst form of extravagance in the end. The tremendous power which a stream acquires in time of high water is seldom properly estimated when it is seen moving tranquilly within its usual bounds. When the day of disaster comes, the trifling neglect which brought it on is regretted, but too late for remedy. Our object in the present volume is to offer a few suggestions which may be of practical value to our readers, in regard to the methods by which a dam may be economically built and yet be of effectual service and the greatest attainable durability. The power of water being by far the cheapest motive power which can be applied to manufacturing purposes, it is important to inquire how it may be employed with the least expense, without sacrifice of strength and utility.

We may perhaps safely venture the assertion that in this volume is presented a larger number and greater variety of dams of different styles, materials and methods of construction than can be found illustrated in like manner in any other work upon the subject; for although a vitally important branch of the business of milling, it has generally been suffered to pass with comparatively slight attention. The building of the mill itself, the application of the motive power and the construction of the machinery for its transmission, are matters upon which innumerable authorities may be consulted; but for the erection of the dam, which may be justly regarded as the very foundation of the enterprise, the mill-owner has been left in a great measure to his own devices, aided by such conflicting counsels as he can obtain from his neighbors and his mill-wright. It has been our aim to supply in some degree this manifest deficiency in the literature of the milling business, and to present the results of a somewhat extended experience and observation in a form which will render them of service to others who are seeking information on this subject.

The Construction of Mill Dams.

CHAPTER I.

MATERIAL AND FORM OF DAMS.

The weirs or dams thrown across the beds of rivers have been constructed in a great variety of shapes and of different materials, some of them too costly for general use in a country where small mills are chiefly needed. In cases where the supply of water is large and a high fall is not demanded, a temporary dam composed of boulder stones is sometimes thrown across the stream in a diagonal or slanting direction, and of length considerably greater than its breadth. The water is thus partly forced into the conduit or race above the dam, and the remainder passes over the surface of the dam in a shallow sheet. Being hastily and cheaply built, a dam of this kind may be repaired without much outlay, but the inconvenience of doing this after every heavy rise of the stream is a material drawback on its value.

In contrast with this comparatively rude species of dam are those of more solid structure, substantially built of stone, and stretched across the river in the form of a bow, the curve being against the current—the middle of the dam, in other words, being higher up the stream than the two ends. A dam of this sort, if provided with massive stone abutments, presents a firm resistance to the onset of a flood, and will stand any test ordinarily experienced. It may be made with a gentle slope from the crest both up and down the stream; or with a steep descent on each side, making its walls almost perpendicular; or again with either a steep or sloping front on the upper side and on the lower a curved apron, the wall rounding downward from the top like the lower half of the letter C, by which arrangement the fall is made gradual and its force abated.

In a stream of moderate size, a form of weir has sometimes been adopted resembling the letter V, with the apex or point directed up-stream. If built upon piles, with a frame of timber forming an inclined plane upon the face of the dam, and filled up with gravel surmounted by a mass of boulder stones well packed in, the dam will be nearly impenetrable by water. The position of the two arms of the V distributes the force of the water in passing over, and as the currents descending from either side tend toward the centre of the stream, the banks are less liable to be washed away. If timber is abundant, the frame, instead of having a uniform slope downward on the face of the dam, may be made in a series of steps like a wide stairway, breaking the water into cascades. The piles for such a dam may be placed at right angles with the current, stayed and covered with plank, and made watertight with sheet piling supported by foot piles. Constructed in other respects like the one last described, a dam of this kind will possess great durability and admit of no leakage.

An undue accumulation of water above the dam may be remedied by a channel and sluice gate in one of the side walls, by which the surplus water may be drawn off before reaching the crest of the dam. A self-adjusting dam of heavy planks strongly framed together is sometimes stretched across the stream, connected by hinges to the crest of the permanent dam, and held in an upright position by weights passing over wheels on the abutments. In case of a flood the weights give way partially to the increased pressure and the auxiliary dam is let down toward a horizontal posiiton, allowing the water to pass unobstructed. In place of an appendage of this kind, movable flash boards are often used, being held in place by pins and other supports along the brink of the dam, and tightly fitted to each other. In time of low water, the flash boards are of important service in obtaining sufficient head. When the stream rises, the boards are removed (though the supports may often remain) and the crest of the main dam being below high water mark, the surplus water escapes freely.

In the following chapters the varieties of dams more practically adapted to the wants of mill-owners in our own country will be mainly considered—including log and frame dams, embankments, crib-work, and their various combinations. We accompany each chapter (with the exception of the 1st and 2d) with a full-page engraving, in order to present to the reader more clearly the suggestions we desire to offer. The methods of construction above described are chiefly useful for large establishments and corporations, with whom the matter of expense is not a vital consideration. Our next inquiry will be how the same practical reliability may be obtained, on a smaller scale and with the most moderate outlay.

CHAPTER II.

MATERIAL AND FORM OF DAMS.—*Continued.*

In many localities where stone is not readily obtained—which is the case in a large portion of the Western States—frame dams are the cheapest substitute, and if properly constructed serve their purpose in the most satisfactory manner. If the stream has a firm and level bottom, the frames, which are made in a triangular shape, resembling a harrow, may be placed directly on the bed of the river, without any intervening foundation. The narrow end or apex is of course laid up the stream, and the frames placed in a line extending from bank to bank, with a space of three or four feet between them. The upper side being then planked the whole length of the dam, an inclined plane is presented to the current on the up stream side, and if the frames are substantially built the pressure of the water will be firmly resisted.

On a soft or irregular bottom, where a heavier foundation is required, the following plan is the most economical, and requires comparatively little labor. Three tiers of timbers running parallel to each other across the stream are placed at the foundation of the dam, one tier at the lower side, one at the upper side, and the third midway between them. Posts are then framed into the lower and middle tiers of timbers, those on the lower tier being of a hight nearly equal to the top of the rafters at the crest of the dam, and the top of those in the middle tier being in range with the former and the foot of the rafters. Two upper tiers of timbers are framed upon the posts thus erected, and the rafters, which should be slightly notched at the point where they rest upon the timbers, are thus firmly supported at the head and foot and in the middle; and the planking being well fastened to them, a strong and serviceable dam is the result, with but moderate outlay either of money or labor.

The best form of dam, whatever be the material of which it is constructed, is that resembling a bow, with the arch up-stream, as described in Chapter I; but this method of construction is seldom followed in frame and log dams, the straight line calling for much less labor and care, as well as less material, the distance being of course less than when a curve is made.

A cheap and substantial dam may be made, where timber is abundant, by laying a foundation of logs of considerable size, which are placed lengthwise of the stream and close together, forming a sort of corduroy road, extending from bank to bank. If the bottom is soft, the logs should be carefully fitted down and adapted to the inequalities of the bed, and if placed as deep as possible they will be less liable to decay by exposure in time of low water. The breastwork of the dam is built near the up-stream side of this foundation, the logs extending from under it down-stream, and serving as an apron to receive the waste water as it comes over. The rafters and coverings of the dam form an inclined plane on the up-stream side, and extend over the upper ends of the logs, protecting the foundation from being undermined by the water working beneath it.

In a region well timbered, and where the stream has a rock or other solid bottom, a log dam of the following description has advantages in point of cheapness, strength and durability. A series of large logs are placed in line, one at the end of another, at the down-stream face of the dam, the loose rubbish being carefully cleaned out, and hollow places filled with short logs to support the main foundation firmly. The logs used for the foundation tier should be as long and large as can conveniently be procured. A series of short logs are then laid upon this tier, with their butt ends resting on the foundation logs and their top ends on the bed of the river, pointing up stream, the distance between them being six or eight feet. Upon these a second tier of long logs is placed, parallel with the foundation tier, but a little farther up stream. A second set of ties is laid with the butt ends on the second tier of logs and the top ends on the ground beside the first set. This second set of ties being a trifle shorter than the first, room is left to place a log of moderate size across the ends of the first ties. This will serve as a support for skids upon which to roll up the third tier of large logs. The logs should be notched where they cross and the ends resting on the ground firmly secured in order to impart the necessary strength to the whole structure. If properly built, the front of the dam will rise considerably faster than the rear, and will at the same time incline up stream, so that its form will resemble a portion of an arch, the foot of the ties being the center and the breast of the dam the circumference. Beside the series of large logs in front, a second and even a third series of smaller size, running parallel with it across the stream, may be placed in the angles formed by the ties, which should be notched where they cross the logs; and the three series of logs should range in hight so that the covering of the dam will form an inclined plane—not too steep for the length of the incline, or the whole fabric may slide down stream when the pressure of the water is brought to bear. Either logs or rafters may be used in constructing the covering. If the former, they should be close together, and chinked with moss, pounded cedar bark, or other suitable material. If rafters are used, they may be placed about three feet apart and planked crossways, the thickest planks being used at the bottom and at the crest of the dam.

Many other varieties of dams remain to be more fully noticed in the succeeding chapters of this work—as every section of country calls for new adaptions of the material and shape of the structure to the peculiar surface, soil and resources of the region. As a matter which may interest the reader, we give, before closing this chapter, a brief account of the method in which a broken dam in the western part of Indiana was repaired some time since, after repeated unsuccessful attempts. The dam referred to was built of logs, brushwood, &c., and the bed of the stream was a treacherous quicksand—perhaps the most difficult kind of bottom upon which to obtain a secure foundation. The breach was nearly in the center of the dam, and not less than forty feet wide; and the deep and swift current which rushed through it defied every effort to gain a point of support from which to work. Various expedients were tried, even to throwing huge boulders into the stream, which were carried along by the force of the water and rendered of no avail. Several millwrights and engineers coasted in skiffs about the place for two or three days, looking for some base of operations, but entirely without success. At length one of them, having pushed his explorations along the banks a mile or more up stream, discovered a huge tree, with an extraordinary breadth of branches, near the water and leaning toward it in such a way as to suggest that it might possibly be launched into the stream. Other trees near it were felled and leaned against standing trees so as to serve as a kind of skids, and by dint of two or three days' patient toil, in which the use of windlasses was found necessary, the tree was at last guided down the stream butt foremost, and lodged in the gap at the broken dam. The expedient proved completely successful, for although a mighty cracking and splintering of boughs ensued, the stout branches held fast on the sides of the dam, and a beginning was thus made from which the necessary repairs were effected with comparative ease.

CHAPTER III.

A BEAVER DAM.

We depart in this chapter from the line of discussion previously laid out so far as to substitute, in our illustration, a dam constructed under the guidance of instinct for one in which human reason and the rules of mechanical science are the directing agencies. The art of dam-building as practised by beavers may be said to possess novelty rather than intrinsic value to the mechanical reader, and we offer the accompanying view of it as matter of interest and amusement to such readers, affording a perhaps not unpleasing variety, in connection with the strictly practical details to which our previous as well as our succeeding chapters are confined. It may be said, indeed, that an examination of the science of constructing mill dams, intended to cover as nearly as possible all the principal branches and phases of the question, would be grossly incomplete if we failed to take notice of the fact that this most useful art has a plain foundation in nature and was practised, in what approaches very closely to an intelligent method, by other animals than man, long before civilization had built its first mill and set its first rude water-wheel in motion. The ingenuity of these untaught workmen furnishes, moreover, a striking argument in favor of the theory, of late a very popular one, that mankind are not a wholly distinct order of beings, but possess a common origin, however remote it may be, with what is conventionally termed the brute creation. This is a matter, however, into which the purpose of the present work does not permit us to enter; especially as it leads to no definite conclusions, but launches the inquirer on a vast field of speculative controversy with no apparent prospect of useful results.

The American Beaver, a specimen of whose skilled labor

A BEAVER DAM.

is shown in our engraving, is remarkable as being the only quadruped known to naturalists whose fore parts resemble those of a land animal, while their hind parts are adapted to aquatic habits, having membranes between the toes. The beavers begin to congregate for their season's work in June or July, on the shores of a lake or river where they design making their abode. They gather from all directions, to the number usually of two or three hundred. They frequently establish themselves in the standing waters of a lake or pond, if of sufficient depth, and in this case no dam is required; but a running stream gives them the advantage of a current by which to transport their supply of wood, which they invariably cut at points farther up the stream than the locality of the dam, being thus enabled to float it down. They use chiefly the trunks and branches of small birch, mulberry, willow or poplar trees, beginning the work of cutting timber early in the summer. The building of the dam and houses is not generally commenced until some time in August, being finished just as the cold weather sets in. The trees are cut so as to fall into the water if possible, although small shrubs at some distance from the stream are dragged to the bank and launched. The strength of the beaver's teeth and the industry he displays are strikingly shown by the size of the trees, reaching sometimes twelve and even eighteen inches diameter, which he cuts down; and this in such numbers and with such regularity that his clearing might easily be mistaken for that of an enterprising settler or backwoodsman.

More surprising still than their sagacity in floating their materials to the proper place is the intelligence shown by the beavers in adapting the shape of their dam to the character of the stream. If the current is gentle, the dam is built nearly straight across; if it flows swiftly, a curved dam is made, arching up-stream. The trees, branches, &c are simply allowed to rest upon the bottom, and mud and stones mixed in, to which the stream itself makes constant additions. The dam thus acquires great solidity and strength, and in some instances the willow or birch trees take root, so that a sort of hedge is at length formed. An old dam which has been often repaired is found to possess remar power of resisting the current and the broken ice d against it.

The main object of the dam is to produce sufficient of water to cover the entrances to the beavers' houses enabling them, in the winter season, to pass in and out the ice. The houses are built of the same material a dams and are very strongly constructed. Each lodge us contains four old beavers and six or eight young ones, th double this number are sometimes found in a single d ing.

CHAPTER IV.

LOG DAM FOR SOFT OR SANDY BOTTOMS.

In a country where timber is abundant, a log dam is the most economical, affording, if properly built, an ample degree of strength and durability at comparatively small cost aside from the labor involved. This is of course a considerable item, but as the work can be performed by ordinary laborers chiefly, it is of a less costly nature in proportion to its amount than the construction of a frame dam or the building of a solid stone wall, requiring the services of a carpenter and a stone-mason.

We give with this chapter an illustration showing with some minuteness the construction of a log-dam which may be relied upon to resist almost any conceivable pressure from the current, and which is in very little danger of being undermined. It is adapted, moreover, to river beds which are of too yielding a character to afford a solid foundation for a stone dam. Our engraving reproduces in all its essential features a drawing lately made in our office for the construction of a dam in the State of Texas. It is adapted to all localities where timber is not too costly, and especially to streams which have soft and sandy bottoms. In the engraving the dam is shown as if cut into at the middle of the stream, the further half being represented, with the crib or pen built into the opposite bank.

The dam here represented may be literally described as a "brush and timber dam," though it comes under the general head of log-dams, the main portion of its structure being of that character, while saplings of any size may be used in making it compact, and brush, clay and boulder stone for filling on the up-stream side. The process of constructing this dam is essentially as follows: Cut trees of eight or ten inches in diameter, lopping off the limbs on what will be the top and bottom sides, when the logs are placed in position. Start the first layer (forming the foundation and front of the apron of the dam, and projecting down stream as shown at the extreme left of the cut) placing the logs side by side with the tops up stream, the lower or butt-ends being about fifty feet below the point where the dam is to be raised. Having completed this, fall back about twenty-five feet and place a second layer of logs side by side as before, the limbs being carefully lopped off on the under and top sides. Having now two layers or courses of logs reaching from side to side of the stream, start a third layer twenty feet back of the second, and carry it across the stream in the same manner as the others. The fourth course, five feet back of the third, completes the series of successive and overlapping tiers of logs constituting the foundation of the dam and forming the apron. With this last course, you begin raising the dam, using for the purpose trees and saplings of any convenient size, and all the while filling in compactly, especially toward the up-stream extremity of the dam, with brush and clay. If boulder stones are readily accessible they should be thrown in along with the clay.

The successive courses of logs should now be laid on in such a way that the face of the dam will present a steep slope, the crest being about two feet farther up-stream than the point at which the dam rests upon the apron. At nearly every course it is well to lay binders lengthwise across the stream, pinning them to the largest logs beneath them. The ends of these binders, which may be three or four inches in diameter, are shown in the cut. They should be placed from two to four feet back of the face of the dam. Having reached the crest of the dam, a top binder is pinned on as solidly as possible, a pin being driven wherever there is a chance for it to hold. If convenient, two or even three binders may be employed, in which case they should be firmly secured to each other and to the upper tier of logs. The dam should be filled in on the surface, from the crest back to the extreme up-stream tips of the trees, with fine brush and clay. For this purpose, trestle work may be built

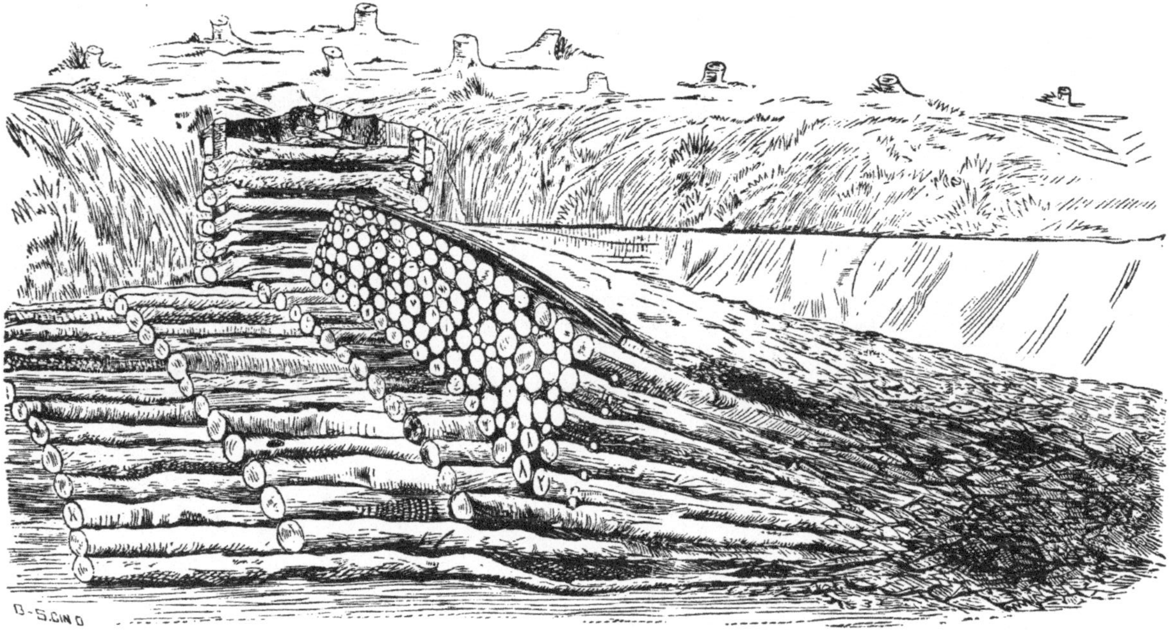

LOG DAM FOR SOFT OR SANDY BOTTOMS.

2

out over the stream and planks laid on to serve as a track on which to wheel the dirt out upon the dam. Throughout the whole work, care should be taken to lop the branches from the top and bottom sides of the trees, and the butts of the trees should invariably be laid down stream. The dam should be made in the form of a semi-circle or half-moon, arching up-stream.

To secure the ends of the dam, a log-pen should be built at each bank, (one of which is shown in the cut,) extending back into the bank as far as it can conveniently be carried. Each pen should be chinked from the inside and filled with clay; or if stone is plenty it may be used instead of clay for filling the pens, which will not then require to be chinked. If clay is used it should be packed in as tightly as possible to prevent it from working out.

It has been found that a dam of this kind will settle about eighteen inches the first year, for which due calculation and allowance should be made. After that time, it will remain nearly stationary. It is cheaper, in a favorable locality, than a frame dam, and has an important advantage in the fact that it will hardly ever wash out. It is almost impossible to build on a quicksand bottom a frame dam that will stay in, as the experience of many mill-owners has shown. The use of piling cannot be recommended, as the water forms small whirlpools around the piles, and will in time wash out the earth clear to their bottoms.

It should be remarked that in building a dam of this kind, unless the stream is nearly dry, it will be well to leave a passage through which the water may escape while the building is going on. This need not usually be done until the apron is completed, and perhaps one or two courses of the upright part of the dam laid on; but after that it will be expedient to leave a space or channel near the middle of the dam for the water to pass through until the rest of the dam is finished, when the gap may be closed up.

CHAPTER V.

A SAFE AND ECONOMICAL DAM.

Although the material and method of construction of a dam should generally be determined by the character of the stream and its bed and the supply of timber or stone to be found in the vicinity, there are many cases in which the natural resources of the country are not specially abundant in either respect, and the most economical method is the best, whether the dam be of logs, frame-work, or stone. It is understood, of course, that strength is under all circumstances the first consideration, and there is no true economy in a plan in which it is not so regarded. The cost of rebuilding or repairing a dam will usually be several times greater than the additional outlay which would have been required to make it proof, in the first place, against sudden floods and the action of frost, floating ice and driftwood.

We give in connection with this chapter an illustration representing a style of dam which can be confidently recommended for its durability, and which involves no excessive outlay either of money or labor. It has some points of resemblance to the log dam described in our last chapter, but the dam referred to is particularly suited to a region where timber is abundant, while the one here shown does not require so ample a supply of that material. The abutments of this dam, it will also be observed, are built of stone, instead of crib-work as in the former case. Crib-work, however, can be substituted if more convenient; or if stone is used, cap-rock or rubble will answer the purpose, if compactly laid and filled in with earth solidly in the rear, nearly as well as more costly building stone.

The construction of this dam is shown very thoroughly in

the cut, and only a brief explanation is necessary. Its qualities of firmness, compactness and durability adapt it to any sort of bottom, whether it be sand, soft mud, gravel or rock; and in either of these cases it would take a more violent pressure than most dams are subjected to, to move it from its foundation. The first step in the process of building it is to lay the foundation logs A A, which extend across the stream, being spliced if necessary, to obtain the requisite length. They should be imbedded in ditches crossing the stream transversely, and of sufficient depth to bring the upper surface of the logs nearly on a level with the bed of the stream. One of these logs is laid at the foot of the apron, another at the point where the dam is to be raised, and the third, fourth and fifth farther up stream, as shown in the cut, the distances between them being six or eight feet. The ends of these logs should project some distance from the sides of the dam into the bank or under the abutment. The weight of the abutment resting upon them will have the effect to hold the dam in its place and prevent it from being lifted or moved forward by the force of the current.

The second series of logs, B B, are laid across the first course, lengthwise of the stream, and about six feet apart, the butt ends resting upon the lower log of the first course. The dam is then raised at the second log of the foundation, which is six or eight feet from the front log, the intervening space being occupied by the apron of the dam. A log of considerable size is laid down directly above the foundation log, and notched to the logs B B wherever they cross. A smaller log may be laid in like manner above the third foundation log, and also at the fourth if desired; as these cross-logs or binders, which should be put in with considerable regularity, especially at the face of the dam and near it (as shown at D D) will serve to support and hold together the whole fabric. The alternate courses of logs having been carried up to the height of about five feet at the face of the dam (which should be nearly perpendicular) and sloping back gradually as shown in the cut, the crossing of the logs will form cribs or chambers, as at C, which are to be filled with stone or gravel. Stone is to be preferred if conveniently at hand, but gravel answers the purpose nearly as well. The binders on the top of the dam, crossing the stream, are to be firmly fastened to the logs upon which they rest—those at the crest of the dam and at the lower edge of the planking being secured by bolts, which pass through all the successive logs below them to the very foundation of the dam. Except at these points it will be sufficient to fasten the binders with long pins to the logs beneath them. The planking E already referred to extends from the crest of the dam about twelve feet toward the up-stream end, and serves to protect the front of the structure from damage by ice or driftwood. The spaces at H, at the rear of the dam, below the planking, should be filled with stone or gravel.

The abutments F (only one of which is here shown, as our engraving comprises but half of the dam, giving a front, top, and sectional view) are built, as already stated, upon the ends of the foundation logs projecting from the sides of the dam, aiding thereby to hold the structure in its place. The abutment represented in the cut is of solid masonry, good building stone being the material employed; but it may be more cheaply constructed, either of rough stone or crib work, as described in the introduction of this chapter.

The apron of this dam should be planked between the projecting logs, as shown at G, the planks extending back under the first transverse log which begins the face of the dam.

It will be manifest from the nature of its construction that no part of this dam can be moved from its place without the entire fabric going with it. The different portions being firmly connected and secured to each other, the structure must go out bodily or not at all. The great breadth of the dam at its base is one of its strongest advantages, preventing it from being undermined by the current—a danger which constantly threatens a dam with a narrow foundation, let it be ever so strongly built. It is also to be observed that the amount of timber required in building by this plan is very moderate, being much less than is often used in

A SAFE AND ECONOMICAL DAM.

dams which do not possess nearly so much actual strength as is here afforded. As a practical example of the reliable character of this dam, we may here remark that one of the publishers of this work, having had two costly dams of cut stone carried away from the same site by high water, finally built one according to the plan here described, at a total cost of $700, and has found it perfectly safe, the floods of four successive years, some of them extremely violent, having failed to carry away any part of it, or inflict any material damage. Neither of the two stone dams which preceded it stood over eighteen months, the bottom being of the sandy and treacherous nature to which a great part of the difficulty involved in the science of dam-building is to be attributed.

CHAPTER VI.

A HOLLOW FRAME DAM.

Having treated at some length on the subject of log-dams. especially for soft or sandy bottoms, and with particular reference to the means at command in a region moderately well timbered, we present in this chapter an illustration of a hollow frame dam which we can safely recommend for a country where economy in timber is necessary. It offers an equal degree of security with the log dams previously described, at the same time requiring far less material in its construction. A good, reliable dam, in fact, can hardly be built with a more economical use of timber than is shown by the plan here given. Our artist has presented very clearly the peculiarities of construction, leaving it necessary to supply little more than the figures, dimensions &c., to give the reader a clear idea of the design. The dam here represented is built upon a solid rock bottom, but with slight modifications is adapted to streams with a soft or sandy bed, as hereafter explained.

The first step in the construction of this dam is to lay the foundation blocks A A, each of which is a stick of timber ten inches square and about four feet in length. Three rows of these blocks are to be laid across the stream, one at the face of the dam, one at the up-stream extremity, and another midway between them. The distance between the centers both ways—across the stream and from one row to another—is eight feet, giving three blocks or bearings to each bent of the frame-work. Three additional blocks are placed in the front row and twelve in the rear row to receive the bolts by which the dam is fastened to the rock. Upon these blocks are now laid the mud-sills B B, which form the immediate foundation of the dam, consisting of logs about sixteen inches in diameter, hewn on the upper and under sides so as to give a thickness in that direction of thirteen inches. These are laid across the bed of the stream in three tiers, one for each row of blocks. Where joints occur, a two-feet splice should be made, and the two ends firmly pinned together. The end of the front sill at each bank should project into the abutment about fifteen inches; while that of the second or middle sill projects an equal distance just behind the up stream wall of the abutment, the center of which is near the front sill, bringing part of the abutment against and the other part below the dam. The front sill has three bolts passing through it, one at each splice, an extra block being placed underneath as already stated. The up-stream sill has twelve bolts, under each of which is a block, in addition to the blocks on which the bents of the frame-work are to rest. The bolts should be $1\frac{3}{8}$ inches in diameter. Each one of them passes through the sill and block down into the rock, which it penetrates about three and a half feet, making the total length of the bolt five and a half feet. The bolt is made after the hole has been drilled, the necessary length being ascertained by careful measurement. A " stoved head " as it is called, is given to the bolt, and a washer placed underneath the head, which is drawn tightly down by the tapered shape of the head. In order to prevent any possibility of the bolt working loose, the lower end is split five or six inches up, and an iron wedge inserted. When the bolt is driven down, the wedge, coming in contact with the rock, is driven up, and spreading the point, holds the bolt firmly in its place. Fine wet sand afterward put in will make it perfectly tight and solid, being as effective for this purpose as lead or cement.

In drilling the hole in the rock, an ordinary stone drill slightly smaller than the hole to be made, is employed, and is put down through the sill and block, which are previously bored and placed in position.

The bents of which the framework of the dam is constructed, and which come next in order, are built throughout of timbers ten inches square—the same size of material being used in the lower horizontal pieces C C, the uprights

D D, and the upper pieces E E, which form the slope of the dam. The length of the lower timbers is sixteen feet, and that of the upper timbers the same, the effect of which is to give the face of the dam a slight inclination up stream. The lower timbers are framed into each sill, a gain being cut two inches deep, and the timbers secured with a dove-tail key driven to the side of each bent. The upright posts connecting the upper and lower timbers of the bent have a length of two feet three inches in the clear at the face of the dam, and half that length at the middle sill; and they are to be mortised into the upper and lower timbers in the same manner as in the framing of a house. The bents are the same distance apart between centers as the blocks under the sills —eight feet, and the distance from the front to the middle upright is the same. The upper and lower timbers of each bent are hewn obliquely or beveled at the up-stream end so as to fit snugly together and give a combined thickness at the extremity equal to one piece.

The last step in the building of the frame is placing the ties F upon the top of the structure, extending transversely across the stream in the same direction as the sills. There are five series of these ties, one over each sill and one between. They should consist of timbers 4 by 7 inches and lie on the narrower side. Each tie is let into the upper or inclined timber of the frame wherever it crosses, the depth of the gain being one and a half inches, giving the tie five and a half inches thickness above the frame. The gain is cut into the frame at right angles with the upper timber, the ties being thus slanted slightly upstream and presenting a level surface for the planking. The forward tie is let into the frame piece about four inches from the end, in order to give sufficient strength to the gain to prevent it from breaking out.

The whole upper surface of the frame is now planked over. The planking, which is strongly spiked to the ties, should be one and a half inches thick, and the wider the better, as the fewer the number of joints, the more secure from leakage will be the covering of the dam. A greater thickness of plank than that given will increase the liability

A HOLLOW FRAME DAM.

to rot, as the wood is wet on one side and dry on the other.

The abutments, as already stated, extend but half way from the face to the up-stream end of the dam. To protect the exposed portion of the sides, the dam is enclosed with stout upright planking from the middle sill to the upstream end, the ends of the planks resting on the rock bottom. In like manner, the rear of the dam is closed with sheet piling extending from bank to bank, closely matched and of sufficient hight to meet the planks which cover the top of the dam, the lower ends of which are footed by the piling, which extends to their upper side and is flush with the surface of the dam.

The abutment is built of timbers fifteen to eighteen inches in diameter and is eleven feet square. The logs are hewed on one side to give a face to the abutment. The first or foundation timbers are laid in the same direction as the sills of the dam, transversely to the stream, the lower one about three feet below the face of the dam, and the upper one just below the middle sill, which it touches and helps to hold in position. The first cross piece on the side toward the dam is laid over and across the end of the front mud-sill, which extends beneath it, as already stated, about fifteen inches into the interior of the crib. The up-stream end of the cross-piece reaches to the middle sill of the dam. The timbers of the crib are notched and saddled where they rest upon each other, and the structure is thus firmly held together. The ends of the first two ties on the surface of the dam extend to the crib, and the third tie passes directly behind it in the same manner as the center sill below. The crib is filled up with rough stone or coarse gravel, and covered with upright planking on the upper side and on the side against the dam. A joist, I, two by ten inches, is spiked against the crib along the top of the dam from its crest to the up-stream corner of the crib.

The dam here shown is ninety-three feet long and its total hight from the rock bottom to the surface of the planking is six and one-half feet. There are eleven bents in the complete dam, only half of which is shown in our engraving. The dam here represented was built by Messrs. Book-

Walter & Claypool of Attica, Ind., to furnish power for a large and very complete flouring-mill just erected by them, in which three Leffel Double Turbines were placed, with all other necessary machinery, furnished by the same establishment. The design of the dam, which was drawn in the office of James Leffel & Co., can be adapted with some alterations to a stream having a soft instead of a rock bottom. For that purpose, it would be necessary to lay a foundation of two and one-half inch plank, instead of the blocks, for the sills to rest upon. These planks should be laid lengthwise of the stream, and project ten or twelve feet below and an equal distance above the dam, making a total distance of about forty feet. As it is difficult to obtain planks of this length, the foundation may be laid in two sections, the planks in each having a length of twenty feet. About midway between the breast and the up stream end of the dam, where, if the planks are twenty feet long, a joint will occur, a wide sill should be placed beneath them, upon which the ends of the planks can be firmly spiked. At the down-stream end of the planks, constituting the edge of the apron, a light sill or binder should be placed underneath—not to support, but rather to hold together the planks. At the up-stream end, the planks will be simply imbedded in the soil, and the planking at this point, and the whole back of the dam, covered with gravel, sand and dirt. A layer of brush at the bottom of this covering will make it hold all the more firmly to the bed of the stream.

CHAPTER VII.

A RIP-RAP DAM.

The conditions of cheapness in the construction of a dam are changed by every change of locality. In one section, where material of a suitable kind may be comparatively abundant, while labor is scarce and commands high wages, economy is consulted by making the work of building as short and simple as possible, even if the material used is not the cheapest which could be found. In another district, or under different circumstances, workmen may be easily obtained at a very moderate rate, and the mill-owner may in this case save money by putting an extra proportion of labor instead of expensive material into his dam. Our engraving herewith given illustrates a kind of dam wholly distinct from any which we have before presented. In some portions of the country it would be difficult to find stone enough for its construction at any price—and it is not intended, of course, for the demands of such l calities. In other sections, the earth and stone of which it is composed would cost almost literally nothing, and it has the further advantage of requiring no skilled labor in putting it up, except in building the chute and waste gate, and in the laying out and general superintendence of the work.

The construction of a "rip-rap dam," which is the term commonly applied to a dam of the description here shown, is begun by throwing an embankment of earth across the stream (space being left in the middle of the stream for the waste-way or chute) carrying it up to a hight of about eight feet in the center and sloping it as shown in the cut, quite steeply on the down-stream and more gradually on the up-stream side. The dam here illustrated has an extent between the foot of the up-stream and that of the down-stream slope, of from thirty to forty feet, and from one bank of the stream to the other of a little over seventy feet. Of the latter distance, twelve feet in the middle of the stream is occupied by the chute (in which the waste-gate is placed as hereafter described) leaving a distance of thirty feet on each side, from the frame-work of the chute to the bank. It is not intended, of course, that the water should at any time flow over a dam of this sort, the escape of surplus water being provided for by the chute. The two slopes of the embankment do not meet at the top in such a way as to form a sharp ridge or crest, but the summit is leveled off so as to give a nearly flat surface about four feet in width, extending the whole length of the dam from each bank until it reaches the chute.

In constructing the embankment, the framework of the chute is to be set in position and strongly planked on the interior side, where the water is to pass, before the earth is filled in at that point—except that a fill about two feet in depth is made on which the floor of the chute is to rest. This floor is laid upon a frame of heavy sills and cross-timbers, the planks of which it is composed extending lengthwise of the stream, and projecting at the down-stream end some eighteen inches beyond the face of the dam, in order that the current of water, as it issues from the chute, may be carried beyond and clear of the embankment beneath. The tendency of the water to wash away the foundation of the dam is thus avoided. This is an important point, as the result of neglect in this particular will be the speedy undermining of the chute and caving in of the wall of earth and rock on either side.

The earth-work having been completed, the dam is now to be "rip-rapped" from end to end. This process consists in laying two courses of stone, one above the other—ordinary cobble-stones being a suitable material for the purpose—over the whole surface of the embankment. The stones are placed on their edges, in the manner in which a gutter is paved, and laid as compactly together as possible, so as to give the entire dam a strong and durable face on both slopes and along the crest. The united depth of the two courses of stone will be about twenty inches. If three instead of two courses

5

A RIP-RAP DAM.

are laid, additional strength will be gained, and the dam will be all the more secure from the effects of any accidental inroad of water. The rip-rapping should not be confined to the dam itself, but extend along the banks on both sides of the stream, a short distance above and below the dam, as shown in the cut. This will prevent the banks from being worn away or washed out, and protect the dam from injury to which it would otherwise be constantly liable.

The up-stream slope of the dam is covered with earth from the base about two feet upward, reaching to the floor of the chute.

Our engraving represents, in addition to the dam, the inlet and part of the channel of the mill-race, on the further bank of the stream. The corners of the banks at the point where the water enters the race should be rip-rapped in the same manner as the dam, to secure them from being washed away and caved in by the continual action of the current. The exact distance to which the sides of the race at this point should be covered with stone will be determined by the shape of the bank, character of the soil, swiftness and force of the current, and other considerations which vary in different localities. The matter will be easily regulated by the exercise of a fair degree of judgment; but in general it is best to err, if at all, on the safe side. A little extra precaution, resulting in perfect security, is better than a falling short which may lead to damage and destruction in time of flood.

The construction of the chute and waste-gate is a matter in which, of course, some measure of skill in the carpenter's and millwright's trade will be in demand. The heavy timbers required are the sills and cross-timbers of the floor, the upright posts, the inclined or slanting beams which follow the direction of the slope of the dam up and down stream, and the timbers connecting them at the top, which will be as long as the crest of the dam is wide. The posts are mortised into the sills below and into the beams above, and their lengths are so arranged as to give the proper slant to the inclined beams, parallel with the face of the dam.

For the construction and operating of the gate, a number

of methods are in use. A very simple arrangement is that in which the gate is raised and lowered by the use of a lever inserted into holes in the standard to which the gate is attached. A chain and windlass may also be used, the manner of their application being so obvious as to require no minute description. Still another form of gate is found very useful, in which the gate is made in sections, each section swinging on a horizontal axle resting on journals near the bottom of the gate, so that it can be let down like the tail-board of a cart when desired, and raised with equal ease whenever necessary. The division of the gate into sections, or as it were into several narrow gates, each acting independently of the other, is found expedient on account of the great force it would be necessary to apply to raise and lower the entire gate in the manner described. The gates fall in the up-stream direction, their own weight assisting the process when they are lowered, and the force of the current helping to raise them—sometimes more powerfully than is desired—when the chute is to be closed.

In future chapters of this work, descriptions and illustrations will be given of one or more varieties of head gates, which with some modification might be applied to a case of the kind above described.

CHAPTER VIII.

A CRIB DAM.

We present in this chapter an illustration of a dam peculiarly adapted to streams which have a comparatively narrow channel, with a high bank on each side—although the latter condition is not indispensable, as any deficiency in this respect, if the shape of the country is not extremely unfavorable, can be made up by constructing an artificial levee or embankment. The structure of this dam is of the nature of crib-work throughout, logs being the material used in every part, although stone, gravel, clay and brush are employed in filling at various points, as hereafter described.

The dimensions of the dam shown in the cut are nearly as follows: length of span, fifty feet, the logs in each of the two sections being about thirty feet long, giving ample margin for notching at each end; cribs on each side twenty feet square, the logs of which they are built from twenty-two to twenty-five feet long and the hight of the cribs from twenty-five to thirty feet. The dam itself is twenty-five feet high, the cribs being carried up three or four logs above the top of the dam.

In building a dam of this description, the whole structure, including both the cribs and the V shaped connection between them, are begun and carried up together. The apron, however, is first put down, consisting of a layer of logs placed closely side by side from bank to bank, with the butt ends down stream, and the limbs lopped off up to the point where the dam is to rest upon the apron. Above this, the limbs may be left on the trunks, as they extend into the earth which is filled in above the dam. The front of the apron should extend three or four feet forward of the cribs, as shown in the cut. The logs used in building the apron, and also the cribs and the dam itself, should be, if possible, at least one foot in diameter, in order to give the proper degree of weight, strength and solidity to the fabric.

Having completed the apron, the next step is to lay the foundation of the wings and central portion of the dam. The first log of the crib on each side should be firmly pinned to the apron; or the foundation of the crib may be laid two or three feet deeper than the apron, in which case it will not be necessary to fasten them together. The cribs are each to be set into the bank, which will thus enclose them on three sides, as appears in our illustration. Thus situated, it is scarcely within the bounds of possibility for the cribs to be moved from their position; and if their connection with the dam is made firm and secure, the strength of the fabric, aided by the peculiar shape which it presents to the current on the up-stream side, will resist almost any conceivable pressure of water.

In building up the cribs and the dam, the logs are to be notched and saddled wherever they meet—that is, at the four corners of each crib, at the points where the timbers of the dam enter the crib, and at the middle of the dam where the two sides of the angle or V intersect. This angle is of course pointed up-stream, the proper distance from the centre or place of intersection to the down-stream edge of the apron being about twenty feet. The pressure of the current upon the harrow-shaped structure thus presented to it will of course tend to spread the two wings or cribs apart; but if the latter are well grounded, filled and supported, and the logs in every part of the dam carefully notched upon each other, the force of the current will have no perceptible effect.

Binders are to be inserted in each half of the dam as the work progresses, one for every second course of logs being sufficient, although one for each course is still more effectual. Small trees or saplings may be used for this purpose with the limbs and brush left on, the butts resting between the logs of the dam and the tops forming a part of the filling on the up-stream side. In the engraving, the ends of these binders may be seen between the courses of logs

A CRIB DAM.

forming the V, the tops of course being covered up and invisible.

The cribs are to be filled with stone and gravel, and if these materials are scarce, a moderate proportion of clay may also be introduced. The up-stream side of the V is to be covered with upright planking, which will extend from the top log down to the apron. Planks ten inches wide and two inches thick are suitable for this purpose, and they should be placed close together and either pinned or spiked to the logs, as convenience may dictate. The planking is cut away at the points where the binders occur, sufficiently to admit the ends of the binders, which rest upon the horizontal logs and are notched to them as already described.

The filling on the up stream side, against the planking, completes the building of the dam. For this purpose, any convenient material may be used, whether stone, gravel, clay or brush, or all together. The filling should slope gradually from the crest of the dam, extending up stream a distance of not less than twenty-five feet, in order that all risk of the washing or undermining of the dam may be avoided.

If the banks of the stream are too low to enclose the cribs to a sufficient hight to make them secure in their position, an artificial embankment must be constructed, covering three sides of the crib and extending from the stream unti it reaches ground of the same hight as the top of the dam. This embankment should be made wide and substantial, and compactly built of stone or earth. It is important that the material should be of such a nature that the water will not penetrate it, as the destruction or serious injury of the dam may occur in consequence of a very small outlet. The main force of the stream is brought to bear, of course, upon the dam itself; but in time of high water there will be more or less pressure upon the levee, which should accordingly be made as secure as circumstances will allow.

Our illustration shows, also, the entrance of the race above the dam, on the left bank of the stream.

The dam above described is adapted to any sort of river-bed, whether it be rock, sand or clay. The shape of the banks is a more material point than the nature of the bottom, especially if it is desired to raise the dam to a equal to that shown in our engraving.

CHAPTER IX.

DAM FOR ROCK BOTTOM.

Under favorable circumstances and with plenty of material of the right kind conveniently at hand, a dam may be built of a character which will do excellent service and withstand the most sweeping freshets, without requiring the exercise of much mechanical skill on the part of the builder. Among this class are several of the dams already illustrated in this work; such, for example, as the log dams shown in our fourth and fifth, and the rip-rap dam in our seventh chapter, all of which might be constructed, with the exception, perhaps, of the chute in the rip-rap dam, with such means and experience in the carpenter's or millwright's trade as are possessed by nearly every mill-owner of average capacity. It often happens, however, that a moderate amount of skill in the use of carpenter's tools and in the putting together of frame-work will enable the builder of a dam to accomplish his object with far less material than he would otherwise be compelled to use, thus reducing very greatly the cost of the undertaking. His head and hands, in other words, will save his pocket, if he is so fortunate as to know how to use them with due effect. For this reason, it is of prime importance to every owner of a water power which is to be utilized by the erection of a dam, to be able to economize and turn to the best possible account the means which nature has placed at his disposal.

Our illustration herewith given presents a description of dam specially adapted to streams which have a solid rock bottom. It is entirely different from any which we have laid before our readers in preceding chapters, and depends for its strength and security mainly upon the thoroughness with which it is put together. With proper care in this regard, intelligently exercised, it may be relied upon as a perfectly safe and permanent structure; this fact having been proved by as severe tests as are generally experienced in the way of sudden and violent floods.

The general construction of this dam is plainly shown in our engraving. The posts A A are supported by the braces B B, and have the horizontal timbers D D gained into them, the upright planking being nailed to the horizontal timbers on the up-stream side. The posts A A should be sixteen inches square, and in the dam here shown are eight feet long above the rock, in addition to which they should extend from twenty-two to twenty-eight inches into the rock, thus giving them a firm foundation and foot-hold. The braces should be twelve inches square, and a 2½ inch shoulder should be made in the post to receive their upper ends, which are beveled to fit the notch in the post. Each brace is secured with a thin key at the top where it encounters the post, and a piece of 1½ inch plank is inserted at the foot of the brace between the end of the brace and the rock. The brace is let into the rock ten inches, the end being left square, and resting in an angular hole or notch ten inches deep and fourteen inches wide.

The horizontal timbers are six by ten inches square, and the gains cut in the posts, in which they are placed, are 2½ inches deep. They may be fastened either by spiking or keying until the upright planking is nailed on. This planking should be two inches thick, and the upper ends of the planks should project one or two inches above the upper horizontal timber or tie. The lower ends should be carefully shaped to fit the uneven surface of the rock with which they come in contact, so that the escape of water beneath them may be prevented as thoroughly as possible.

The upper part of each post is beveled in the down-stream direction to enable the water to run off freely, and also to prevent it from going down the grain of the wood and thus causing it to decay.

The holes for the posts are drilled in the rock in dovetail shape on the upper side. The end of the post is cut to fit this dovetail, a shoulder being made on that side two inches

DAM FOR ROCK BOTTOM.

deep. For the admission of the post it will be necessary, of course, to make the hole somewhat larger than the end of the post. For the purpose of securing the post firmly in its place, a long, wide key, about 2½ inches thick, is inserted on the lower side of the post, holding it snugly and strongly against the dovetail side of the hole. This key should be so made as to fit the full depth of the hole against the rock on the lower side. It is of vital importance that the dovetail should be made on the up stream side. Experience has shown in the most striking manner the necessity of this arrangement. A dam in one instance within our knowledge was built on the general plan here described, but the dovetail and key placed in contrary order to that here given—the dovetail being on the down-stream side and the key put in on the up-stream side. This dam went out when a flood came upon it, the keys, which perhaps were not fitted with sufficient accuracy, working loose so that the posts lifted out. When re-built in the same manner, except that the dovetail was placed on the upper and the keys on the lower side, the dam stood firm, resisting up to the present time every freshet that has occurred.

The length of span of the dam here illustrated is about sixty feet, the distance between the posts being twelve feet.

On the up-stream side of the dam a fill should be made against the planking, extending up from the rock four or five feet and sloping off gradually. For this fill coarse gravel or small stones may be first used, immediately against the bottom of the planks, finer material being afterward thrown in.

For the upright planking, it is well to use seasoned and green planks alternately. If all are seasoned, they are liable to swell in such a manner as to burst them off of the ties. If all are green, exposure to the air will cause them to shrink and leave wide cracks. The planks should be champered or half-beveled on the up-stream edge (the down-stream edges being brought together perfectly close) in order to catch the loose earth which may drift against them.

On the left of our engraving is the race H, and an elevated wing of the dam, the post G being about five feet er than those in the main portion of the dam. The c extends into the rock on that side of the stream a there firmly secured. The planking F is nailed to thi below; and in this wing of the dam is the head-gate, does not appear in the cut, but the position of wh clearly indicated.

CHAPTER X.

A PILE DAM.

The dam which we illustrate in this chapter is adapted to a mud bottom or to any kind of river-bed which will afford a firm foothold for piles, and into which they can be driven to the necessary depth. The first step in the process is the preparation of the piles, which should be of oak if convenient, ten or twelve inches in diameter, and from twelve to twenty feet in length, according to the hight it is intended to give the dam, the nature of the bottom, and consequently the depth to which it is necessary to drive the piles. The taper at the lower end should begin two or two-and-one-half feet from the point. In using the pile-driver in setting the piles it will be found that the force of the successive blows will after a time have the effect to split the pile at or near the top; and to prevent this a ring should be placed over the top of the post. This ring is made of bar iron three-fourths of an inch thick and from two and a half to three inches wide, the ends where they meet to form the circle being welded as strongly as possible. It is also expedient to champer or bevel the inside of the ring so that it will go on with the wider opening downwards, the object being to make the ring compress the top of the pile in such a manner that when it is desired to remove the ring it will come off easily. Care must be taken not to champer it at too great an angle, or the post will act upon it like a wedge and the ring will burst before the driving is completed. As this accident is liable to happen in spite of every precaution that can be taken, it is well to have several rings made before beginning the work, so that while one is taken away for repair another may be used and the driving go on without interruption.

For a dam like the one here illustrated, the piles are driven eight or nine feet into the ground, leaving from six to eight feet above for hight of dam; but where the bottom is sound and firm the posts need not go in to so great a depth. There are three rows of piles shown in our engraving, the two front rows, A and B, being close together, but alternating so as to "break joints," and the second and third rows, B and C, being far enough apart to admit a horizontal layer of logs E E between them. A dam may be built with but two or even only one row of piling, and possess sufficient strength for any ordinary test. If only one row is planted, logs and brush should be piled up behind it on the up-stream side so as to make the dam tight and break the immediate pressure of the current. The horizontal logs E E should be of about the same diameter as the piles; and between them, at intervals, are inserted the butt ends of the binders G G, which are logs or poles from thirty to forty feet in length, extending from the piles up stream and being covered with the filling. The upper horizontal log E is pinned, as will be seen in the cut, to the end of the binder below it; and this should be done at frequent points along the whole extent of the span.

The apron H has a foundation of heavy sills D D, for which large logs should be selected, laid transversely across the stream, and spliced and firmly pinned where two ends meet. Cross logs F F are laid upon these, extending up-stream between the piles, and having a length of from ten to fifteen feet, or whatever may be necessary to prevent the water, as it comes over the dam, from striking beyond the apron and washing out the river bed. The planks H are laid parallel with the cross-logs F and firmly spiked to the sills D D.

The crib I is a hollow square composed of piles driven down in the same manner as those of the dam, but of greater length above the bed of the stream, making the top of the crib from two to four feet higher than the dam, according to the hight and nature of the bank. These piles, as will be seen, are placed close together; and the dam should meet the crib at a point a little farther up-stream than the center of the crib. At K is shown the water-line when the dam is full to the crest; and at L is indicated the filling, for which gravel, dirt and stones may be used, the slope extending

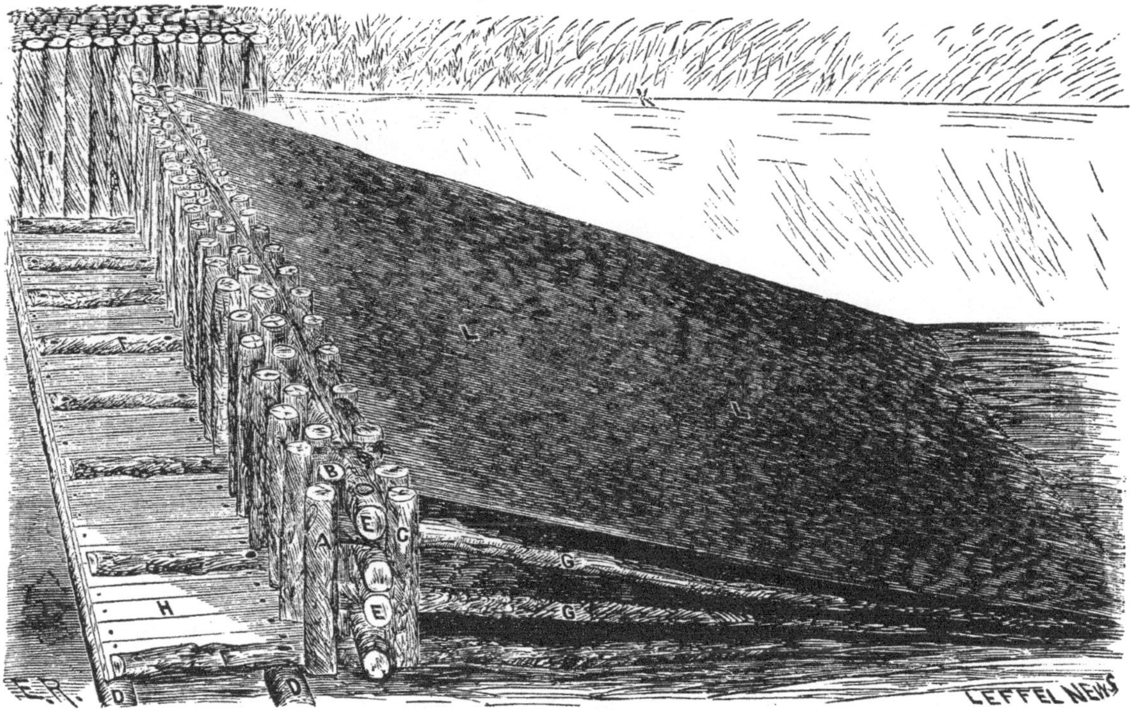

A PILE DAM.

from the crest of the dam thirty or forty feet up-stream The same material is used for the filling of the cribs or abutments, of which but one is shown in our engraving, which represents the dam as if cut in two, lengthwise of the stream.

It is of special importance that the dam should be made as nearly water-tight as possible in every part. For this purpose stones may be used in filling the holes, the size of those first put in being sufficient to prevent their passing through, and smaller stones being thrown in after these. The same process is of equal benefit in the interior of the crib, where there is considerable liability of washing out. Hazel brush cut fine and closely packed in is also recommended for this purpose, the only objection being that it is subject to decay in course of time.

Upon a bottom of suitable character, a dam built upon this plan will be found very substantial and reliable. It is of course not adapted to localities where quicksands occur at the bed of streams, as the piling cannot in such a case be given a firm foothold.

CHAPTER XI.

THE HOUSATONIC DAM AT BIRMINGHAM, CONN.

Our suggestions on the subject of dam-building have thus far been chiefly confined to enterprises of a comparatively limited scope, being within the means of a single mill-owner of moderate capital. As this class comprises nineteen-twentieths, at least, of all the persons immediately interested in the use of water-power, it is of course entitled to the larger share of attention, and it is for the benefit of such readers that this volume is principally designed. As a variation, however, of the general plan thus far adhered to, a description of one of the most extensive works of this nature ever carried to successful completion will possess sufficient interest to reward an attentive perusal. Such an enterprise is the erection of the Dam of the Ousatonic Water Company, which extends across the Housatonic River at Birmingham, Connecticut, and which was some three years since brought to successful completion. There are but few instances of a work of this kind conducted on so large a scale, and involving so immense an increase of manufacturing facilities.

The damming of the Housatonic River was a subject of discussion as long ago as the year 1838, and at about that time a petition was presented to the Legislature of Connecticut having this object in view. Only a low tumbling dam, however, was permitted to be built, a high one being forbidden on account of its preventing the passage of the shad—a higher value being then attached to the shad fisheries than to the manufacturing interest. This compelled, also, a change of the location of the dam, and ultimately it was found that half a million dollars would be required for its erection, whereupon the enterprise was abandoned. It did not again take practical shape until the year 1863, at

THE HOUSATONIC DAM, BIRMINGHAM, CONN.

which time the movement was inaugurated which has since been crowned with such triumphant success. The interval from 1863 to 1867 was consumed in financial and legislative preliminaries, negotiations for real estate, obtaining of the charter and capital stock, &c., the Company being organized in November, 1866. The plans and specifications were then made by Wm. E. Worthen of New York; Henry T. Potter was engaged as Engineer and Superintendent, and the first stone was laid July 17, 1867. In August of that year, and again in September, the work was interrupted by freshets. The double difficulty arising from the current of the river above and the tide below with its three feet rise and fall every twelve hours, rendered the laying of the foundation an arduous task; and it was found necessary to build coffer dams of plank, backed with earth, the water being then pumped out of the enclosed space. One of these dams was broken by the August freshet, but was speedily repaired. The stratum of rock at the bed of the river was found to dip too much to admit of the masonry being united with it, and the foundation was therefore laid on the gravel above, into which sheet piling was driven, the ends projecting upward a few feet and being encased in the stone-work. In November, four months from the commencement, about 200 feet of the dam had been built, some twelve feet above the foundation.

In 1868, the 200 feet of dam in progress the previous year was completed and 300 feet more of foundation laid, leaving a gap of 100 feet for the passage of the stream, when a freshet swept through all the coffer dam protecting the unfinished work. This was late in the season, and the coffer dam was not restored until the spring of 1869, when it was nearly carried off a second time by a June freshet which caused two weeks delay in the enterprise. The work then progressed until the gap in the center had been nearly closed, the water having been turned through the headgates on the west side, when the heaviest disaster of all occurred, on the occasion of the great freshet of Oct. 4th. The central portion of the dam had been left in the worst possible condition, the back being carried up several feet higher than the front. The water passed over this part thirteen feet deep, carrying away the coffer above and

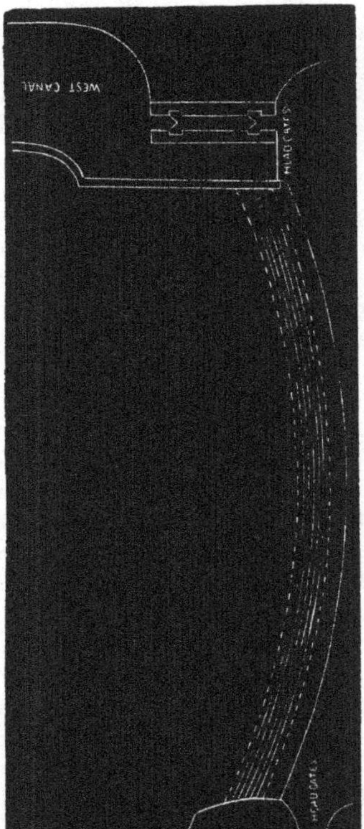

HOUSATONIC DAM—(Ground Plan.)

undermining and sweeping away about 160 feet of the dam. No attempt was made at rebuilding until the following year, when the coffers above and below the dam were restored, the latter being finished early in July. The removal of the water from the immense coffer below the dam was a work of such magnitude that the engineer, Mr. Potter, devised a pump expressly for the purpose, 48 feet long, 4 feet wide and 12 inches high, with buckets or elevators attached to belts. The power for this huge elevator was furnished by a turbine wheel, enclosed in a large frame-work built on the apron at the west end of the dam, and using a portion of the water from the river then flowing through the head-gates, to conduct which to the wheel a temporary flume was constructed. Geared to this wheel was a large fly-wheel 12 feet in diameter, and another 190 feet distant, driven by a rope-belt above the pump. At each end of the pump was a drum around which the elevator-belts passed, and into which one-half of each bucket fitted. This enormous pump worked night and day for several weeks, throwing about 5,000 gallons of water per minute, and enabling the workmen to pursue their task below the level of the river. When the water had been all removed from the coffer, it was found that the full extent of the damage done by the October freshet had not been realized. It had not only swept away the central part of the dam but had cut down the river-bed south of the dam, making a hole more than half an acre in extent and 20 feet deep below the apron. This immense cavity was filled with rock and stones, the foundations laid upon it, and on the 5th of October, 1870, the last coping stone was laid. On the 14th of that month the water was running over the dam, and the enterprise was an assured success.

The Housatonic dam, thus completed in spite of the most formidable obstacles, is of solid masonry, 870 feet long including the abutments, and 22 feet 6 inches high. The curve which it makes as shown in our second cut, constitutes the arc of a circle having a radius of 2,000 feet. The base is 20 feet, and the front has a slope up stream of 2½ inches per foot. It is capped with blocks of Maine granite 8 feet long and one foot thick. The whole structure, including the surroundings, is estimated to contain 451,000 cubic feet of masonry. For the protection of the base from undermining an apron is provided, as indicated in our third cut, 24 feet long, composed of timber and concrete, and having 10-inch

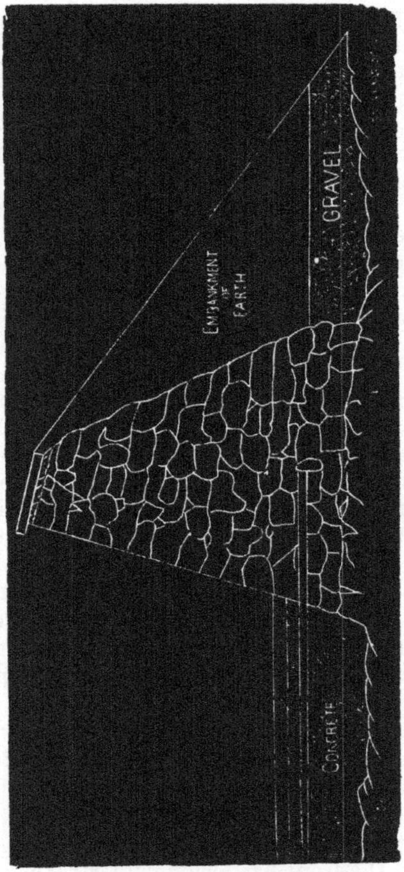

HOUSATONIC DAM—(Vertical Section.)

sills extending 8 feet into the stone-work of the dam. These sills are imbedded in concrete, and a second course of timbers of the same thickness are bolted to them at right angles. All the spaces are filled with concrete, and the surface of the apron is then laid, consisting of timbers a foot square, lying close together in the same direction as the lower sills, and strongly bolted to the timbers underneath.

On each side of the river is a canal conveying the water for use by the factories. That on the west side, which is the larger of the two, has five gateways each eight feet square, with solid pillars of stone two feet thick between them, the bottom laid in cement and the top slabs of stone. The gates are made of oak planks 8 by 8 inches, strongly bolted together and shutting in grooves in the stone. The canal is 60 feet wide and 14 feet deep, giving a cross section of 840 square feet, and has an overflow of 150 feet near the dam. The sides are walled with stone. In its complete state it will open 3,700 feet of factory front, or over two-thirds of a mile.

In June, 1871, the discharge of the river was ascertained to be nearly 5,000 cubic feet per second. Early in the fall the discharge was still one-fifth the above amount, in the midst of a drought which stopped nearly all establishments dependent upon water-power. The mimimum average flow at the lowest stage of water is estimated at 500 cubic feet per second, equivalent, with the head of 22 feet, to 2,500 horse-power for 12 hours per day. With the ample reservoir, extending back five miles and covering nearly one thousand acres, no apprehension can exist of a lack of water.

The almost unequalled extent of this water-power, combined with its close proximity to New York, with which it is in direct communication by water, in addition to ample railroad facilities in all directions, have given it wide celebrity, the advantages which it offers being apparent at a glance. The principal office of the Ousatonic Water Company is at Birmingham, Conn. E. N. Shelton, Esq., is President, and associated with him in the management are Messrs. D. W. Plumb, Dr. John I. Howe, Edwin Wooster, Robert N. Bassett, A. H. Alling, R. M. Bassett, Wm. E. Downs, and Thos. Elmes. D. S. Brinsmade, Jr., a graduate

of the Scientific Department of Yale College, is
gineer now in charge.

CHAPTER XII.

A PLANK CRIB DAM.

We return in this chapter to a class of dams more generally interesting to the milling community, because more applicable to the circumstances and wants of the vast majority, than such extensive structures as that described in our last chapter. For every locality where capital by the hundred thousand or half million dollars can be judiciously invested in a dam, or is at command for such a purpose, there are hundreds where a small water-power offers a profitable business to a single operator with limited capital, and where the best method of utilizing that power becomes a question of vital interest. For the benefit of this numerous class of persons we illustrate herewith a dam which offers peculiar advantages in the manner of its construction, admitting at once of a high degree of strength and very close economy in point of material. Its ability to resist the force and weight of the current is founded upon one of the simplest principles of mechanical science—that of the arch, which is employed in so many forms in the builder's art where a heavy load is to be sustained or a powerful strain endured. It is hardly possible to conceive of a pressure, except it were an upheaval from below which should not occur from any other cause than an earthquake, by which a dam built in the manner here illustrated could be forced from its position. We base this statement, of course. on the supposition that due care is exercised in the construction of every part, and especially, as will be hereafter indicated, that the crib at each end of the dam is solidly built and firmly imbedded in the bank.

Our engraving shows the dam as if cut in two in the middle of the stream, the other half being exactly similar in construction, subject to such modifications as the shape of the bank may require.

Upon a rock bottom, no foundation sills are needed; but on a soft bottom, these must be first laid, lengthwise of the stream, and consisting of logs 10 or 12 inches in diameter, with a flat face hewn on the upper side, reducing the vertical thickness to about 8 inches. The length of these logs will depend upon the extent of apron it is intended the dam shall have. Upon a mud bottom a twelve foot apron will be sufficient; but upon quicksands an apron thirty or forty feet long will be required to prevent washing out under the front of the dam; and in this case, adding the apron and dam together, the logs should be about sixty feet long, while with a twelve-feet apron logs thirty or forty feet long will be sufficient. They may be laid about two feet apart on a mud bottom, but upon a sandy bed should be placed close together, the two edges being straightened so as to admit of a snug contact. If laid in this manner on a mud or stony bottom, the apron, consisting of the ends of the logs projecting from beneath the front wall of the dam, will not require to be planked; but on a sandy bed, or in case the two-feet spaces are left between the logs, the apron must be planked as shown in the cut. The foundation sill is represented in the engraving as a timber accurately squared on all four sides; but this is not necessary, as a log with the flat upper face and two straight edges will answer every purpose.

The foundation having been completed, reaching from bank to bank, the walls of the dam, the two cribs and the apron are next to be constructed. The cribs and the dam are carried up together, and the planks of which the apron is composed (the cross-sills having been previously gained upon the foundation logs and firmly pinned) are extended into the front wall of the dam as will be seen in the cut. The sills for the apron may consist of timbers six inches square, and 2½ or 3 inch hard wood planks, strongly pinned to the sills, should be used for the covering at this point. For the walls of the dam and cribs, two-inch planks are sufficient; and every other plank in the wall of the dam

A PLANK CRIB DAM.

should pass through that of the crib, thus constituting a part of the partition wall inside. The effect of this arrangement will be that half the planks in the dam will pass into the crib, and the other half will abut against it, so that it will hold firmly against the pressure brought to bear on it, from whatever direction it may come, and no separation of the dam from the crib can in reasonable possibility occur.

Before carrying up the walls, however, it is necessary to fix the direction of the arch or up-stream curve of the dam. For this purpose a general rule may be given, as follows: Supposing the width of the stream between the two cribs to be 100 feet, take a rope of that length and attach it to a stake in the middle of the stream far enough below the dam so that the up-stream end of the rope will just reach the middle of the inner wall of either crib, or the point where the dam is to rest against the crib; then, keeping the rope tightly stretched and carrying the up-stream end across from one crib to the other, it will describe the curve which the dam should follow. In other words, the front wall of the dam should constitute the arc of a circle of which the radius is the distance in a straight line between the cribs. This is, as we have said, a good general rule for the purpose, though it is subject to variation according to the shape of the banks, depth and force of the current, and other circumstances. Against a very powerful current, or if the banks are low and not very substantial, the dam should have a greater curve up-stream than if the current is moderate, or the banks rocky and firm.

The direction which the two walls of the dam are to take having thus been ascertained—the lower or up-stream tier of planks being parallel with the higher tier—the planks are laid up so as to form a solid barrier across the stream, breaking joints, and each plank strongly pinned to those below it, and neatly fitted and jointed where they meet, and also where they come in contact with the crib. The hight of the front wall, in such a dam as that here illustrated, is about ten feet, that of the rear wall a little over five feet, and the distance between their centers ten to twelve feet. The cribs, composed of two-inch planks, laid up in the same manner, have a partition dividing them in the middle as shown in the cut; and the front wall of the dam should abut against and connect with the end of this partition, as already described. The hight of the cribs will be determined by the shape of the banks, in which they should be imbedded as firmly as possible, being surrounded on three sides by solid ground or substantial filling of gravel or stones.

For the covering of the dam three-inch planks may be employed, placed snugly together and solidly pinned to the two walls of plank on which they rest.

The dam is filled between the walls, back of the rear wall and over the lower part of the covering, with earth, gravel or coarse stone; and the same material may be used to fill the two apartments of each crib.

It will be manifest upon a moment's consideration, that the pressure of the water upon this dam will be like that of the superstructure of a building upon the arch on which it rests, tending to spread the arch outward. As it is held in this case by the crib at either end against which it abuts, it cannot spread out except by absolutely crushing the cribs or pushing them into the bank, neither of which events can happen if the cribs are properly filled and backed up. We do not know of any arrangement comprising so small an amount of material and so simply constructed, by which a greater power of resistance is afforded.

A dam may be built on this principle with still less material, by erecting but a single wall and letting the covering extend back from the top of this wall to the up-stream end of the dam; or the covering itself may be dispensed with, and the gravel and stone simply filled in against the single front wall, constructed as already described. This is the simplest and cheapest form in which the dam can be built. The cribs must be strongly put up, whatever may be the plan of the dam. as the pressure which would tend to spread the dam must in any case be resisted at this point.

A dam of this description may have three walls instead of two, in case it is desired to carry it to a greater hight than that here indicated. Another and more radical change of the plan is to exactly reverse the dam, making what is here the up-stream the down-stream side, the high wall being farthest up-stream and the lower walls below it. In this case the covering extends from the highest or rear wall to the foot of the dam down-stream, making it, in effect, a long inclined apron. For this style of dam, a fill is made on the up-stream side of the high wall reaching about two-thirds the hight of the wall, and extending up-stream far enough to cover the upper ends of the foundation logs. These logs project down stream a short distance beyond the foot of the dam, and may be planked at this point, unless they are laid close together, in which case the planking will not be necessary.

The pins used in securing the planks should $\frac{3}{4}$ of an inch square in a $\frac{3}{4}$ inch hole, where soft wood is used; but in hard wood the pin should be somewhat smaller.

For the covering of the dam, instead of using three-inch planks, a double covering may be made of two-inch planks, laid so as to break joints; or in place of either, what would be called among backwoodsmen a "puncheon floor" may be constructed; a log being sawed or split for this purpose into strips or slabs four or five inches thick, and these pieces scotched at the ends to an even thickness, where they are pinned to the two walls of the dam.

One of the main advantages of this dam is the ease and rapidity with which it may be put up. A hundred men, if necessary, may be employed upon it at once and the work thus carried forward at any speed which the circumstances may render desirable.

CHAPTER XIII.

THE MOLINE DAM.

One of the most extensive and liberally developed water-powers in the United States is that located at the town of Moline, Ill., a place of some 6,000 inhabitants, situated on the east bank of the Mississippi River, immediately opposite the head of the island known as Rock Island, situated about 300 miles above St. Louis and midway between that city and St. Paul. The water-power lies between the Illinois shore and the island, and is near the foot of the upper rapids of the Mississippi—a succession of rapids or falls, extending over twenty miles of the river channel, and having an aggregate decline of eighteen feet in that distance. The effective head is secured by extending a wing wall from the point of the island, nearly three thousand feet in length, which, with the island and main shores, gives a water surface of several hundred acres. With this length of wall, a permanent head of seven feet is obtained, and the body of available water is so large that this gauge can scarcely be perceptibly decreased. The circumstances of the case, also, are such that by simply extending this wall the power can be indefinitely increased; and it is estimated that an expenditure of $30,000 on the work would double its capacity.

An effort was made as early as the year 1843 to develop the power so manifestly and abundantly available at this point; and a rude dam was thrown across the channel and a mill operated with the force thus obtained. Although of slight importance in itself, this enterprise was the foundation of the immense development of manufacturing resources which has since been witnessed, inasmuch as it served to attract a considerable number of settlers to the locality, and in fact gave to the town the name it has since borne—the word Moline being a corruption of the French term MOULIN,

THE MOLINE DAM.

signifying a mill. Improvements were afterward made upon the power, and additional manufacturing establishments erected upon it; but the lack of sufficient capital for such expensive undertakings prevented the work from being carried on either so rapidly or on so large a scale as would have been otherwise attempted. In process of time the power passed into the hands of a company, who held title in connection with it to some two hundred acres of land bordering on and adjacent to the river.

At the close of the war the United States Government, which is proprietor of the island of Rock Island, selected it as the site of a great arsenal and armory for the manufacture and storage of war material, the buildings for which are now in process of erection. In order to avail itself of the Moline Water Power for running the immense amount of machinery to be used in its workshops, the government, in August, 1867, while President Grant was acting Secretary of War, entered into a contract with the Water Power Company, by which the company ceded to the government their portion of the power, the government, on the other hand, binding itself to make certain specified improvements, develop the power at its own cost, and give the company a perpetual title to one-fourth of the whole, free from rent, repairs and expense of every kind whatever. The government also contracted to rent additional power to the company, at a fixed price; so that, with the growth of manufactures, the power may eventually become a source of revenue both to the government and the company.

In pursuance of this contract, the necessary appropriations were made by Congress, and the work was begun forthwith. The old dam was torn out, the reservoir deepened, and the adjoining banks of the island raised and rip-rapped with rock. A wall of the heaviest Joliet rock was built longitudinally with the channel of the river, 2,400 feet in length, twenty feet in hight, eight feet wide at the base and sloping to four feet at top, with supporting buttresses of three feet at intervals of ten feet. This wall is a massive piece of masonry, set in the bed rock of the river, beyond the power of flood or ice ever to disturb it. The adjoining bank, as already stated, was heavily rip-rapped to secure it agains washing; and all the work done by the goverment was of the mos

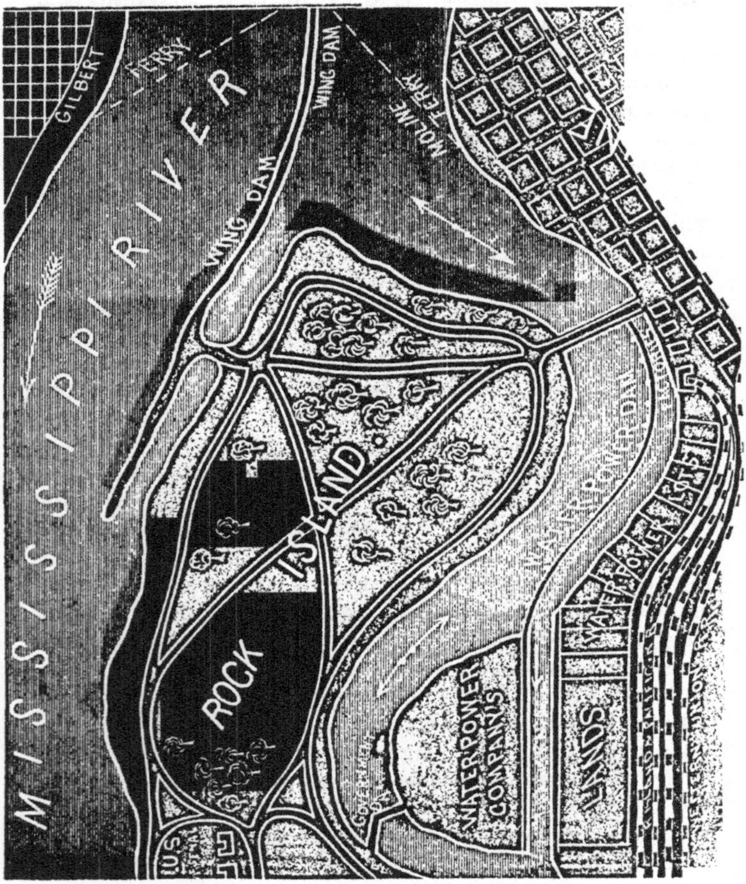

MAP OF THE MOLINE WATER POWER AND VICINITY.

substantial character. Meanwhile, the Water Power Compan) deepened the channel for carrying off the tail water and put in bulk-head; and the manufacturers located on the Power set thei wheels, all of which are now in operation. The work which remains to be done, and in which the government is now engaged, is an addition of 1,400 feet to the longitudinal wall, and the excavation of a canal across a projecting point of land 2,100 feet in length, to carry off the tail water and discharge it below the Power.

The company owns half a mile of river frontage, every portion of which is well adapted for the location of workshops, and provided with ample railroad facilities. The permanent character of the work, and the nature of the contract with the government, are such that parties purchasing mill-sites are secured against any extra assessments for repairs or improvements of any sort whatever. The immense reservoir from which the power is drawn, with its ample and constant supply, renders impossible any perceptible reduction of its head.

In the first of the two engravings herewith presented, a general view is given of the dam and water power, and in the second a full and accurate map of the same, showing the Water Power Dam, the Government Dam, the wing dam, the town of Moline, the factories, water-power lots and lines of railway, Rock Island, the Mississippi River and both its banks. By the aid of these illustrations the construction of this immense work will be clearly understood, and some idea of its magnitude will also be gained.

The officers of the Moline Water Power Co. are Charles Atkinson, President, C. H. Deere, Vice President, and John M. Gould, Secretary, the address of each being Moline, Ill.

CHAPTER XIV.

A BOULDER WING DAM.

There are numerous localities, especially in the Western States, where a water power of considerable value may be obtained by diverting a portion of the current of a stream of too great width and volume for the erection of an ordinary dam, directing the water into the race, and making the latter of considerable length so that a fall of several feet may be produced, with an abundant and unfailing supply of water. In such cases, it has been found in practice a very expedient course to build what is called a Wing Dam, an example of which is presented in the accompanying engraving. This consists of a dam in two distinct sections, one on each side of the stream, and one section considerably farther down stream than the other. The obvious purpose of this arrangement is to throw the current, by means of the upper wing of the dam, to the opposite side of the stream, where it encounters the other wing and a part of it passes into the race, the remainder escaping around the inner end of this portion of the dam. This, of course, does not constitute an economical appropriation of the full power of the water contained in the stream, and is only adapted to cases in which there is a plentiful supply of water. Another and very important use which this description of dam is made to serve is that of forming a channel for navigation in streams which in their natural state are too shallow. An instance of this kind may be observed on the Mississippi river in the vicinity of Keokuk, Iowa, where a rapid occurs in the stream for a considerable distance, and the depth of water, without some artificial concentration, would be insufficient to allow a boat of any considerable draft to pass. A series of wing dams or breakwaters has therefore been constructed at this point, alternating from bank to bank,

A BOULDER WING DAM.

and keeping the water in a comparatively narrow channel, of such depth as to admit of navigation.

A wing-dam is particularly adapted, or rather is most conveniently built, across a stream having a gravel or stone bottom, as its construction renders it peculiarly liable to be undermined at the inner end of each wing. The distance to which the wings extend into the stream will be determined by the width and depth of the river, the strength of the current, and other attendant circumstances. In very wide and shallow streams where the current is not powerful the wings may lap beyond the point at which they would meet if started at directly opposite points; but in a deeper and swifter current this would not be advisable, as the water would be very apt to undermine the exposed ends or abutments of the wings. For the same reason it is a work of somewhat difficult nature to build a dam of this kind upon a soft bottom, especially where quicksands occur; the great obstacle being the tendency of the current to wash out beneath the foundation of the cribs. This can only be obviated by driving down piles very thickly and to a considerable depth, and attaching the timbers of the dam to them. The extent to which the precautions of this nature must be carried will depend upon the character of the river bed and all the other conditions above alluded to, which must be carefully considered and estimated in order that the structure may be made capable of resisting the attacks which will be made upon it.

On a solid bottom, this description of dam may be built in the style and of the material of any of those which have been illustrated in preceding chapters, either logs, planks, rib-work, gravel or boulders being used, as may be found most economical or convenient. In our illustration a dam is shown, constructed of boulders with a crib abutment at the inner end of each wing. The process of building such a dam is very simple, and requires but brief explanation. The cribs are first to be laid, as great difficulty would be experienced in their construction if deferred until the other portions were finished, on account of the current which would then have been created by the confinement of the stream. In building the cribs, logs ten or twelve inches in diameter and about twenty feet in length should be used, notched upon each other and firmly pinned together; and in some cases, even on a moderately firm bottom, it may be well to drive down piles at the corners of the cribs and pin the logs to them as strongly as possible. Upon a rock bottom the foundation logs should be fastened down by means of anchor-bolts. The cribs are of triangular form, the point being up-stream; and they are to be filled with coarse stone or gravel, or any material which is not liable to be washed out. The boulders for the remaining part of each wing are then thrown in, a broad base being given to the dam to secure the requisite stability. The hight of the dam and cribs should be such as to bring them a few feet above low water mark. In time of flood, the water will of course pour over the whole structure.

The construction of the race does not differ from that employed in connection with an ordinary dam, except that to obtain the necessary fall it requires to be carried an unusually long distance below the point at which the water enters it. In some localities the race is made a mile or more in length, the descent being so gradual that an effective head cannot otherwise be obtained. But notwithstanding the apparent wastefulness of this method of utilizing the power of a stream, it is often very profitably employed, and is indeed in many cases the only practicable means of turning to useful account the resources offered by a water course of great width and very gradual fall, or upon which, owing to its interference with navigation, a dam cannot be constructed.

The length of race represented in our illustration is less than will usually be required with a dam of this description, the fall here shown in the stream being comparatively rapid.

CHAPTER XV.

A BRUSH, STONE AND GRAVEL DAM.

We have illustrated in previous chapters dams of many different descriptions, the particular variety represented being in nearly every case adapted to a certain class of localities—a log or crib dam, for instance, to a region where timber is abundant, a frame dam to a stream with a rock bottom, a pile dam to a mud bottom, and so on; the presumption being in these cases that the locality in which the dam is built has either some specially convenient and plentiful material for its construction, or presents such conditions, in the character of the stream and its banks, as to indicate clearly the kind of dam which it will be necessary to adopt. There are, however, many sections of our country in which the selection of a plan for the erection of a dam and of the material to be used in the work is not so easily made—nature not having pointed out the course to be pursued, either by providing means and resources of a particular kind, or limiting the choice of the mill-owner by the necessities of the case, the nature of the river-bed or other attendant circumstances. It frequently happens that a dam is to be built in a locality where neither timber nor rock is extremely abundant, although both are to be had in moderate quantities without excessive cost; and where the bed of the stream is such that any one of a dozen different methods may be followed in the erection of the dam, neither possessing any striking advantages over the rest. In such a case as this it is a matter of economy to the mill-owner to use all the different resources at his command without any disproportionate tax upon either; and by availing himself of all the favorable conditions presented, he can generally make a strong and reliable dam without employing, to any great extent, the skilled labor of the carpenter. A dam of this composite character, including logs, brush, stone, gravel, sand, loam, and even clay in its materials, and depending upon its shape rather than on any peculiarity of construction for the necessary durability, can in many places be made more cheaply than any which we have yet described.

The engraving herewith presented gives a view, taken from nature, of a dam of the kind above referred to, the locality being on Mad River in Clarke County, Ohio, and the owners of the power Messrs. Snyder & Bro., proprietors of a flouring-mill and several other manufacturing establishments. The bottom of the stream here shown is a mixture of mud, sand and gravel, with a low bank of black soil. In constructing the dam, the first step taken was to throw in large quantities of brush, which was piled up until it reached, as it lay in its loose state, a hight of ten feet or more. Boulders and coarse stone were then thrown in, crushing down the brush, and toward the top of the dam finer rock and gravel were put in. The brush and stones, being thus piled and mixed together, had the effect to hold each other in place; and it should be observed that the brush was of all sizes, trees and saplings, some of them forty feet in length, being laid in with the butts down stream.

In topping off the dam, the rocks and gravel were thrown on so as to form a natural slope on the face or down-stream side. The dam was so built as to form a curve, arching up-stream, so as to throw the water passing over it toward the center and thus protect the banks from washing. The length of the dam here illustrated is about 100 yards and the hight about four feet, the stream being of considerable width and comparatively shallow. The crest of the dam is of course irregularly proportioned, but it has an average breadth of about six feet.

Especial care must be taken in putting in the "filling" of a dam of this description—for which purpose gravel, sand and loam are used—to close up thoroughly all the spaces and apertures between the rocks and among the brush and logs. If these are not compactly filled, the water may find

BRUSH, STONE AND GRAVEL DAM.

its way into the interior of the dam and it will be almost impossible to repair the mischief when discovered. If clay is used at all in filling, it must be in small quantity, and thoroughly mixed with the other materials, as it is the most unreliable of them all in resisting the inroads of the current.

The base of a dam of this kind requires to be of considerable extent in order that it may be perfectly durable. A width of twenty-five feet from the foot of the upper to that of the lower slope will in ordinary cases be sufficient. It will be seen by the engraving that a dam of this sort becomes in process of time a permanent barrier to the current, apparently more the work of nature than of art, but none the less effectual in rendering available the power of the stream.

Nearly all the dams on Mad River are of the same general description as the one here represented, and have proved very durable, having stood the test of many years' use, and numerous floods of great violence. The locality shown in our sketch possesses peculiar historic interest, being within a few hundred yards of the birth-place of the celebrated Indian warrior, Tecumseh, the tragical end of whose career has formed the theme of much vivid description and supplied one of the most familiar illustrations of the school geographies of the last generation.

CHAPTER XVI.

CURVED PLANK DAM FOR ROCK BOTTOM.

The great secret of utility in mechanical devices is to derive from a proper application of simple mechanical principles the strength and force which must otherwise be obtained by the use of a large amount of material and of cumbrous and costly appliances and methods of construction. Numerous instances of this kind will occur to the reader, in which "the brain is made to save the hands," or in other words an exercise of judgment and ingenuity in turning the laws of nature to account enables a scientific workman to accomplish his object with but a fraction of the outlay of time and labor which would be required if the work were not thus facilitated. There is perhaps no elementary principle which can thus be made available with more useful effect than that of the arch, which is so constantly employed as an indispensable feature in the construction of bridges and buildings of every class. The stability and sustaining power which are afforded by this form of construction could not in some cases be obtained with ten-fold the amount of timber and stone work put together in any other way; and it is in all cases a source of advantage and of economy in the structure too important to be overlooked. We present in this chapter a representation of a dam in the construction of which the vital and characteristic principle is that of the arch, the strength of the fabric depending mainly on this peculiarity, and so great a saving of material being made in consequence that we may fairly pronounce this the most simple and economical form of dam which we have yet presented to our readers. There are many other structures of this kind, it is true, built with a curve up-stream, mainly for the purpose of directing the flow of the water toward the middle of the channel as it pours over the dam, and thus saving the banks from being washed; but in those cases the dam derives its strength from other sources and is but

CURVED PLANK DAM FOR ROCK BOTTOM.

little affected in that respect by the direction in which it crosses the stream. There is an exception to this in the case of the dam illustrated in Chapter XII, which is built on the same principle as that here shown; but being intended for a sand or mud bottom as well as rock, it has foundation sills, an apron and considerable filling on the up-stream side—all of which are here dispensed with except that a small fill may be made of gravel and dirt against the dam on the up-stream side, to keep it perfectly tight. There is nothing in this dam, practically speaking, but a tight wall of planks laid flatwise, the wall curved up-stream, and snugly let into the rock at the ends—the banks, as well as the bed of the stream, being supposed to consist of rock. A more simple plan could hardly be devised.

In building this dam any kind of timber may be used, as the cheapest will answer the purpose. The planks should be not more than twelve feet in length or thereabouts, as this length is well adapted to making the necessary curve without having to spring the plank. The width should be from 10 to 12 inches and the thickness 2 to 2½ inches. For the first or bottom course a level place should be made in the rock so that the foundation will be even and firm. A good square shoulder must also be cut in the rock in each bank, extending vertically from the bottom to the top of the dam, and of such shape that the square ends of the planks (which will strike the bank in a slanting direction on account of the curve of the dam) will fit into it snugly and closely; and care must be taken in laying the planks at the extremities of the dam to set their ends squarely into this shoulder. When so placed there is no danger of their slipping out or being carried past the shoulder, as the whole pressure of the water against the dam tends to force them into the bank.

The curve which the dam should make will depend somewhat on circumstances, and especially on the depth and volume of the stream. The rule given in Chapter XII was to take a rope of a length equal to the distance between the banks, fasten one end of it in the middle of the stream just far enough below the dam so that the other end would reach the point of contact of the dam with either bank; then, stretching the rope tight, and carrying the upper end across the stream, it would describe the curve which the dam should make, and which would be, in fact, an arc of a circle having a radius equal to the width of the stream. This will answer for the present description of dam; although it should be borne in mind that the greater the curve the more firmly it will resist the pressure to which it is subjected, and any variation from the rule above stated should be on the side of safety.

The first three or four courses of plank having been laid, they are to be bolted to the rock bottom, the bolts passing through each plank and extending from 12 to 20 inches into the rock, according to the nature and solidity of the latter. The bolts should be wedged at the bottom, and the heads countersunk, either the plank containing the head of the bolt or that directly above it being cut to admit the head. The method of wedging at the bottom of the bolt is substantially as follows: the lower end of the bolt is split five or six inches up, and an iron wedge inserted. When the bolt is driven down, the wedge, coming in contact with the rock at the bottom of the hole, is driven up, spreading the point and holding the bolt firmly in its place. Its security may be still further increased by putting fine wet sand in the hole after the bolt has been driven down. As the planks will lie somewhat loosely at first, and will undergo heavy pressure from the water when the dam is used, they should be subjected to pressure in the same direction by means of keys before the holes are laid off and the bolts driven down.

The remaining or upper courses of planks are laid on as indicated in the cut, and secured with spikes or wooden pins, particular care being taken to make every joint snug and close. The dam represented in the cut is about 8 feet high and 60 or 70 feet in length; but the same rules of construction may be followed for a dam of any other dimensions within a reasonable limit, and without any other protection or support than that here shown, which is altogether in the arch or curve of the dam, the rock bottom and the recesses or shoulders in the banks of the stream. It will be plainly seen that a timber dam cannot possibly be built with a smaller quantity of material and possess the same proportionate amount of strength.

CHAPTER XVII.

CONSTRUCTION OF A DAM BETWEEN COFFERS.

The process of erection of a dam which we here illustrate is given not only for the information of our readers at large, but also as a general response to numerous letters of inquiry which we have received in regard to the method of building a dam in a heavy stream, with a deep or very swift current. The difficulties encountered in constructing a dam under these circumstances are of such magnitude as to justify us in devoting to the matter an entire chapter of our series. The necessity of a coffer dam to protect the permanent dam while in process of construction is sufficiently understood by most persons; but the manner in which the work is undertaken and carried through is not so fully known.

The coffer dam here represented is of the kind adapted to streams having a mud, clay or sandy bottom into which the piling can be driven. Where the stream has a rock bottom a different mode of procedure is called for, of which we have only space in this chapter to say that it requires in the first place the sinking of cribs filled with stone, at suitable distances apart, against which crib sills are laid and planks thrust down vertically on the outer side of the sills, against which they will be held by the pressure of the water; or if the cribs are very close to each other the planks may be put down horizontally, the water holding them against the cribs. For the coffer illustrated in our engraving, the first step is to drive down the piling, for which purpose planks should be used eight or ten inches in width and two or three inches thick, according to the depth and resulting pressure of water. For each coffer—the one above and the other below the point where the dam is to be built—two rows of piling are driven down, from two to four feet apart, and to a

CONSTRUCTION OF DAM BETWEEN COFFERS.

sufficient depth to give them strength and firmness, the requisite depth depending on the solidity of the bottom. Particular care must be taken in driving down the piling to keep the planks as close together as possible, so that any liability to leakage will be avoided. The space between the two rows of piling is filled in with any convenient material which will not wash out. Clay will answer the purpose, but soil or loam containing but little clay is to be preferred, and is still better if mixed with hay, straw or long manure next to the sides of the planks. The liability of clay to wash out has been alluded to in previous chapters, and where the other materials mentioned can be obtained it is advisable to use them. Previous to the filling, however, binders must be put in as shown in the cut, for which purpose four or six inch scantling may be used. The cross pieces, of the same material, are pinned to the binders and serve to hold the two rows of piling against the tendency of the filling to spread them apart.

The distance between the upper and lower coffer will be in ordinary cases from 50 to 200 feet, according to the width it is intended to give to the dam. It is important that plenty of room should be left between the coffer and the dam to admit of working and hauling in material from the shore side. If the coffer, owing to very strong pressure or some accidental imperfection in its construction, is found liable to cave in, it may be braced on the inner side, the props extending from just beneath the binder to the ground; or the same object may be accomplished by placing timbers across the entire width between the coffers (if the distance be not too great) above the top of the dam.

The coffers having been carried to the middle of the stream, they are to be connected at that end by a structure of the same kind extending from one coffer to the other, and thus leaving the current of the river to pass on the unobstructed side of the stream. The construction of this end-coffer, which has its direction parallel with the stream, is in all essential respects the same as that of the upper and lower side-coffers. The next step, the entire coffer for this half of the dam having been completed, is to pump the water from the inside of the enclosure, which must be done with a steam engine and pump of suitable capacity. The first half of the dam is then put up to its proper hight. This having been done, an additional end-coffer is put in, within the enclosure, ten or fifteen feet from the mid-stream end of the dam, and parallel with the stream and the end-coffer previously erected. This extra coffer extends between and connects the two main upper and lower coffers, and unites at its middle with the dam, its connections with which must be made perfectly tight. The upper and lower coffers, from the shore to the cross-coffer last constructed, are now torn out, and similar coffers are built from the other side of the stream in the manner already described, connecting with the portions of the first coffers which remain standing. The end-coffer first built is now taken out, which leaves the enclosure on this side complete, with the dam projecting into it ten or fifteen feet. The water is now pumped out of the new coffer and the remaining half of the dam is then built.

The second coffer must be made somewhat higher than the first, as the water, which is changed to the side of the stream from which it was turned by the first coffer, has now to pass over the dam, and will consequently rise to a greater hight on the up-stream side. In the cut, a comparatively sluggish stream is shown, the water being at nearly the same hight above and below the coffer; but in a rapid stream, the water will be lower below the coffer, and the down-stream side of the coffer need not therefore be made so high as the up-stream side.

The dam having been completed in every respect, all the coffers which remain standing may either be taken out or left standing, as desired, and the work is done.

In our illustration, a frame dam is shown in process of construction between the coffers; but it is of course a matter depending upon the choice of the builder and the circumstances of the case what kind of a dam shall be erected. As the frame dam which we have chosen to illustrate in this instance is of a somewhat different character from those represented in previous chapters, we will briefly describe its mode of construction. The sills of this dam are thirty feet in length and 12 inches square. The bents, each one of which is composed of a sill, three upright timbers and two rafters, are placed at a distance of 10 or 12 feet apart, and all the timbers used in them should be of nearly the same width and thickness as the sills. The rafters and uprights are strongly pinned to each other and to the sills. The ribs or stringers of the frame are 12 by 8 or 10 inches, and are bolted to the rafters. The planks of which the covering of the dam is composed are 2 or 3 inches in thickness and from 10 to 16 inches wide, and are spiked or pinned to the ribs. The joints of the planking should be made very close to prevent the dam from leaking; or what is still better, the planking may be made double, the upper course breaking joints with the lower; in which case the planks need not be so thick. In the cut, a narrow apron is shown on the down-stream side of the dam, the planks of which extend inward under the covering, and are spiked or pinned to sills beneath them running lengthwise of the dam. This apron may be made much wider if the flow of water and the nature of the stream render it desirable. The hight of the dam in our illustration is about ten feet. For the filling, gravel, small stones or boulders, mixed with soil, should be used. Upon a tolerably firm bottom this dam will be found very safe and durable, its broad base and the manner in which the water strikes and escapes from it being such as to give it a high degree of resisting power.

In taking down the coffers, it should be here remarked, the filling which has been used in them may be thrown in on the up-stream side of the dam, where it will serve a very useful purpose in preventing the entrance of water and undermining of the structure.

CHAPTER XVIII.

STONE DAM NEAR FRANKFORT, KY.

The science of dam-building is a comprehensive one, examples embraced in it ranging from the rudest cheapest barrier which can be built of earth and stones a small mill, upon a shallow stream, up to the most ex sive and costly structures, such as the Ousatonic and Mo dams illustrated in previous chapters. The engraving h with presented gives a view of a dam which may be ‹ sidered as belonging to the last mentioned class, inasm as it could be profitably built only where a large busine. to be done and an ample amount of capital is inves The massive and substantial character of this dam strike the reader at a glance; and it has, in fact, quit imposing architectural effect with the two towers at its tremities and the heavy masonry extending from and necting them.

The stream on which this dam is built has a rock bott which is leveled down to receive the foundation sto The stones in the main portion of the dam and in the tov are each about two feet thick, four or five feet wide, from six to eight feet long. They are laid lengthwise v the dam, and are cut wedging or beveled at the ends s to fit snugly to each other and form in a solid manner curve of the dam, in the same way that the stones of arch are shaped to correspond with the form of the st ture. They are laid in hydraulic cement, or "water li the nature and use of which substance, the peculiar tr ment demanded for stone-work thus put up, and other tails of this part of the work, belong to the special provi of the stone-mason and will be sufficiently understood competent workman at that trade.

The face of the dam consists of one tier of stones of

STONE DAM NEAR FRANKFORT, KY.

size and built up in the manner above described, and the back or up-stream side is composed of loose flat stones laid up without much regularity, the work being finished by filling with earth and small stones, making a tolerably gradual slope on the up-stream side and giving the dam a wide and substantial base. The face of the dam, composed of the solid tier of masonry, inclines slightly up-stream, varying but a foot or two from a perpendicular.

The length of the dam here shown is about 300 feet between the towers. Its width at the base is 8 feet and at the top 5 feet, and its hight about 12 feet. The towers are each 17 feet high, 18 feet square at the base, and 12 feet square at the top. The number of courses of stone in the main dam is six, the stones being, as already stated, two feet thick.

From the tower at the right of the picture a wing or continuation of the dam runs out, keeping the same general direction as the dam itself, its purpose being to keep the earth from washing out, the bank being common soil and very low on this side of the stream. On the other side the bank is of sufficient hight to require no protection of this kind. The dimensions of the wing wall are less than those of the main dam, as it has much less pressure to resist. From the same tower with which the wing wall connects, a side wall of somewhat greater width and hight extends up the stream, along the bank to the entrance of the race, protecting the bank from being overflowed or worn away by the stream. A similar wall also extends a short distance above the race, and both sides of the race are in like manner strengthened against the action of the water, their walls being somewhat lower than those along the banks of the stream. At the entrance of the race are placed head-gates, which are not represented in our engraving.

A tablet on the face of one of the towers bears the name of the owner of the dam, date of construction, &c.

Our illustration gives a view of the dam by which the power is furnished for the extensive flouring mill of Mr. Geo. B. Macklin, four miles from the city of Frankfort, Ky. The course of the stream at this point is such (making what may be called a "horse-shoe" bend) that by running race from 200 to 300 yards a fall of 21 feet is obtained. ' firm of James Leffel & Co. have put in three of their proved Double Turbine wheels in this mill, from which an idea may be formed of the amount of business it is signed to maintain, and which fully justifies the thoro and durable manner in which the dam has been construc The dam was completed about four years since, and is of the most substantial structures of the kind to be fo in the West or South.

CHAPTER XIX.

AN IRON DAM.

Iron is now so common a material for building, and enters so largely into the construction of edifices of every class, that it would be a matter of surprise if it did not prove adaptable as well to the erection of dams. It possesses the advantages of great strength in small compass and of being more conveniently handled than the cumbrous materials required in building many other classes of dams. We illustrate in this chapter the first dam in our whole series thus far of which iron has formed a part otherwise than in the shape of bolts or spikes. We do not now present it as possessing practical merit superior to that of many other dams already represented, but rather as an interesting novelty, and an example of what may be done in the erection of a substantial and reliable dam, with iron as the almost exclusive material.

The dam shown in our engraving is particularly suited to a rock bottom, or can be placed on a solid earth bottom by putting under it a substantial foundation of timbers. It consists, as to its general construction, of upright sections or panels of boiler or cast-iron, built in zig-zag or "worm" fashion, after the style of the "Virginia fence," and supported at each bank by stone towers. The hight of the panels is six feet, in the dam here represented, and the horizontal length of each the same, so that the distance between the extreme ends of the two panels forming an angle will be ten or eleven feet, the width of the "worm" being three feet. The total length, therefore, of the dam shown in our cut is about 120 feet. Its course across the stream is straight, there being no advantage gained, in a dam of this construction, by giving it the arch or curve formation which has been shown in previous chapters.

The panels, as already stated, may be made either of boiler-iron or cast-iron. If of boiler-iron, each panel is composed of two plates of equal size, lapped and riveted together. If of cast-iron, three plates are used in each panel, filleted together and ribbed on the back. The panels in this case should be about one inch thick. The posts to which the panels are bolted may be of either cast or wrought iron, and are shaped to suit the angle made by the corners. They should be about one inch thick, and contain recesses to receive the ends of the plates. At the bottom of the post is a flange or foot-pad, through which the bolts are driven by which the post is secured to the rock.

The braces supporting the posts are of wrought iron. They are bolted to the posts on the down-stream side, and have a cast flange at bottom through which they are bolted to the rock in the same manner as the posts. The brace should meet the post within six or eight inches of the top of the latter.

The stone towers shown in the cut at the extremities of the dam are about 12 feet high, 12 to 14 feet square at the base, and 10 to 12 feet square at the top. The relative dimensions and manner of construction of the walls enclosing the race will be sufficiently understood from an examination of the cut. The race is provided at its entrance with a head gate of the usual description, made in two parts for convenience of operation.

If a protection is required for the dam, on account of ice or driftwood passing over, a timber may be laid lengthwise along the top of the dam, and rafters placed against it, which when sheeted over will form a sort of inclined apron The ice or other substances calculated to injure the dam if allowed to pass over with a perpendicular fall, will thus be safely conveyed away on a gradual slope. As the "worm" of the dam is three feet wide, as already stated, three timbers instead of one may be laid along the top, placed snugly side by side and constituting a still more reliable covering. This portion of the work may be varied in many ways, ac-

AN IRON DAM.

cording to the judgment or preference of the builder of the dam, and will of course depend greatly upon the peculiar circumstances of the case, the nature of the stream and the country, and the materials most readily at command. In many localities, the dam itself may be all that is required, no covering or shield being necessary if it is subject only to the pressure and spill of water.

It should also be remarked that this is but one of a great diversity of iron dams which it would be practicable to construct, and any one of which would doubtless be equally serviceable with this. In fact, if iron is the sole or chief material employed, it may be put up in any shape or on any system the builder may select, it being in this respect more adaptable to circumstances than any other material. Stone can only be used for laying in a wall, filling cribs, or rip-rapping an embankment; and logs and timbers are subject to a variety of limitations arising principally from the fact that they possess the necessary strength only when in considerable bulk. With iron, on the other hand, the maximum of strength may be attained with the minimum of dimensions. It affords more liberty in the selection of plans and mode of construction than the more bulky substances which are generally used, and which, it must be admitted, have undeniable advantages on the score of cheapness and the ease with which they can often be turned to account without the employment of skilled labor or expensive manufacturing processes.

CHAPTER XX.

PILE AND BOULDER DAMS.

A point of vital importance in the construction of a dam is to provide against the effects of the overflow, which has a constant tendency to undermine the foundation of the dam by gradually washing out or wearing away the bed of the stream at the point where it strikes. This is in fact one of the chief difficulties to be taken into account in building a dam, and it is at this point, perhaps, as frequently as any other which can be named, that they prove to be defective. The working of the water in the manner described is often so gradual and insidious that its effect is unobserved until made apparent by the giving way of the foundation at one or more points, and this very probably at a season of high water, when even an attempt to repair the mischief is scarcely possible and is apt, if made, to prove ineffectual. The location of the breach, moreover, being at the very base of the dam, where the weight of the structure and the pressure of the water bearing upon it is chiefly felt, renders the task of making good a defect of this kind after it has thus betrayed itself an exceedingly arduous one. In many instances it has been found impossible to perform the work successfully, and the loss and entire rebuilding of the dam has proved the necessary consequence of the neglect to make it in the first place secure against this particular danger. In this matter, as in a thousand others, experience has repeatedly shown that an ounce of prevention is worth a pound of cure. A due regard beforehand for the known conditions of the case and the well understood action of water in flowing over a dam will save many hundreds of dollars in subsequent repairs and rebuilding, as well as in

FIG. I.

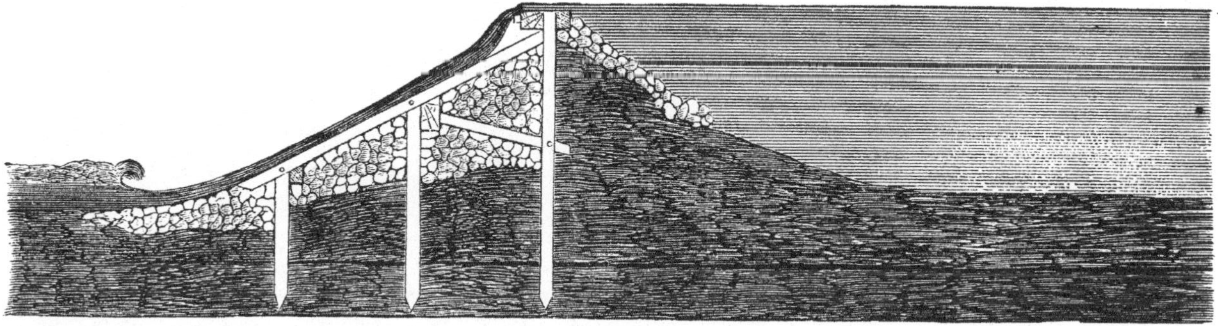

FIG. 2

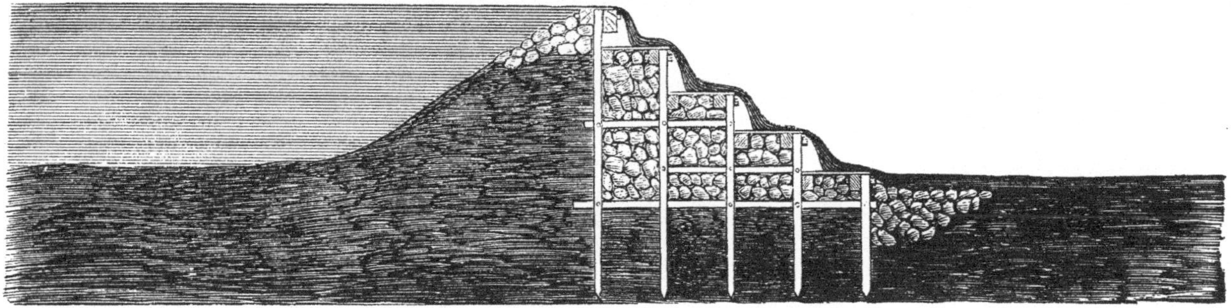

PILE AND BOULDER DAMS.

the incidental losses which attend a breakdown of this kind.

The methods of construction adopted to prevent this destructive action of the water are various, and some of them have already been illustrated in this volume. In some cases the horizontal apron projecting from the foot of the dam down stream is found sufficient; while in others an apron is made forming a gradual inclined plane from the crest of the dam to a point some distance down stream, thus permitting the water to escape in a swift current instead of a heavy perpendicular fall. This is an excellent method of breaking its force, provided the inclined plane is itself strongly constructed with a firm foundation and with no liability to leakage.

We illustrate in the engraving here presented two forms of construction for a dam by which the prevention of injury by underwash is effectually attained. A glance at either of the two figures in the cut will show that there is scarcely any possibility of the water wearing either downward in the bed of the stream or backward under the foundations. Both these methods of construction partake of the nature of the inclined plane, it being so modified in the second as to constitute a series of steps by which the force of the water is broken and it is carried in successive easy stages from the top to the bottom. The strength of the materials used and of the method in which they are employed will also be observed. In Figure 1 an open frame of timber is represented, into which are inserted large boulder stones, forming a compact mass of boulder sheeting resting on gravel, and nearly impervious to water. The timbers here used should be large, say a foot square, and firmly pinned together. The hight of the dam in Figure 1 is about eight feet, and the bents, each consisting, as shown, of the three uprights, the inclined rafter and the intermediate brace, should be placed at intervals of six or eight feet along the face of the dam. A stringer is bolted to the top of the second upright, under the rafter, as shown in the cut, and one end of the brace is let into this stringer, the other end being pinned to the highest upright. At the top of this upright, also, will be seen a heavy stringer bolted on each side in such a way as to form a level surface at the crest of the dam and protect the structure at that point. In addition to the brace shown in the engraving others may be put in in a similar manner, giving additional strength to the framework.

In Figure 2, the same principle of construction will be observed, with modifications suited to a region where timber is plentiful. This structure is composed of piles driven at right angles with the direction of the stream and placed in rows, properly stayed and covered with planking firmly nailed to the horizontal and vertical timbers. If it is desired to have the structure perfectly water-tight, a line of sheet piling may be driven in, in the line of the dam across the whole breadth of the stream, and this being again supported by foot piles and stays at different distances, a tight and very durable dam is the result. The water falls in cascades over the series of steps, and any injurious effect on the foundations is prevented. The method in which the horizontal timbers are pinned to the uprights and the stringers bolted to the top of each upright will be readily understood by a glance at the cut. The timbers in this dam need not be so large as in Figure 1, six or eight inches square being amply sufficient. The bents in this, as in the other dam, may be put in at intervals of six or eight feet.

The construction of both these dams, aside from the framework, is very simple, and presents a safe and substantial resistance to the pressure of the water. The rip-rap of the embankment on the upper side may be carried farther, if desired, than is shown in the cut; and the proportions and extent of the layers of boulders may be varied in any direction according to the judgment of the builder and the circumstances of the case.

CHAPTER XXI.

STONE DAMS.

Whatever may be said in favor of other descriptions of dams, whether they be frame, crib, log, pile, earth, brush or iron dams, it must still be admitted that stone is on many accounts the most suitable material for a barrier against the pressure of water, and the one which will naturally be selected where the circumstances do not make it too costly, or where the object in view cannot be as effectually accomplished by more convenient methods. Stone possesses more of the qualities which are valuable in a dam than any other substance. Its weight, though it renders the work of building more arduous, is a source of strength when it is once in position, such as can hardly be given to any other material; it is subject to neither rot nor rust, and unless undermined or caved, in consequence of the weakness of some other part of the structure, it is not liable to yield to any of the ordinary forces which a dam is intended to resist. When properly guarded from the gradual inroads of the water through apertures or crevices, or in the form of underwash by which the foundations are sapped, a stone dam is an immovable bulwark and will withstand the heaviest freshets, saving in the long run, in many cases, by the avoidance of any outlay for repairs, many times the difference between its cost and that of cheaper but less reliable structures.

We illustrate in this chapter two different forms of a stone dam, both of which have the attributes of strength and permanency, the choice between them depending on the character of the stream and the means of the builder. Figure 1 is a sectional view of a dam constructed in England by the celebrated engineer Smeaton. It has, as will be seen, a long inclined slope on each side, above and below, and the extent of base in proportion to the hight is such as to con tribute immensely to the stability of the dam. At the crest of the dam and inclined downward, will be observed the ends of two courses of flag stones, which are so laid as to break joints, footed upon grooved sheet piling with bearing piles and stringer, and supported by a thick and wide layer of "rubble" or small boulders underneath the flags. Live moss is packed between the flag stones to prevent the silt being driven through. At the foot of the dam is another row of sheet piling, similarly supported, and protected by a fir plank at top from the action of the water. Over the layer of rubble is placed a row of regular stones, laid endwise and leaning, so as to be perfectly secure from derangements by floods. In the up-stream direction from the crest of the dam is also placed a layer of rubble with a tier of flat stones above it and at the up stream end of the dam is a single course of large flags reaching from the surface to the base of the dam and held down by a filling of small stones. A dam of this kind is adapted to a stream having a soft bottom either of loam or clay; but if there is a sandy bottom the piles must be driven to such depth as will give them a firm hold, and special care must be taken to guard against underwash at the foot or down-stream extremity of the dam.

In Figure 2 is shown a dam built upon a somewhat different principle, composed of solid masonry instead of rubble and flags, and having a curved or concave apron instead of a gradual slope on the down-stream side. On the upstream side the construction is very similar to that of the dam in Figure 1, except that stones of more regular shape are used; and in all parts of this dam reliance is placed upon heavy and compact stonework instead of piling. The water has at first a nearly perpendicular fall, but is carried away in such a manner by the curve of the apron as to entirely lose the direction which would give it an injurious effect upon the foundation. The construction of this dam throughout is so distinctly shown by the sectional view here given that it requires no further explanation.

An English writer on the subject of the construction of

FIG. 1

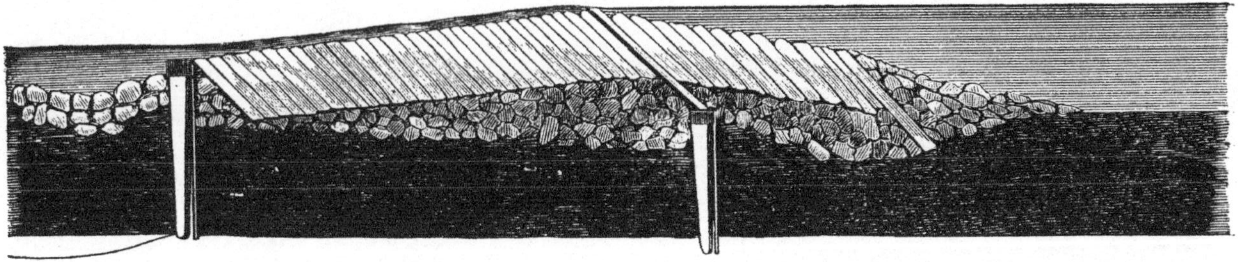

FIG. 2

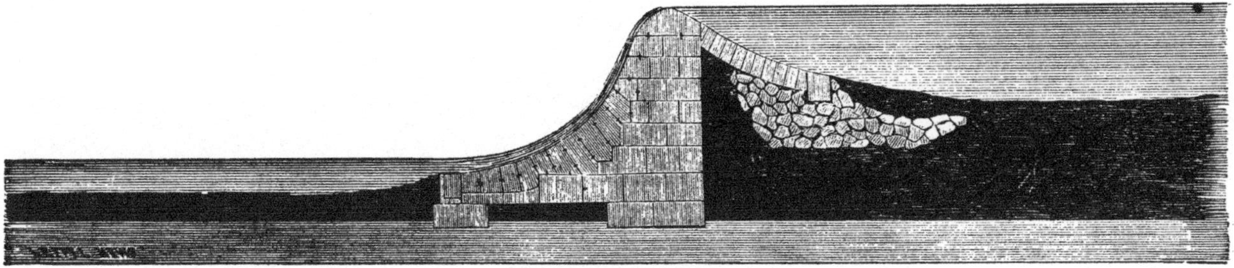

STONE DAMS.

dams remarks that "rapid rising of the waters and sudden changes in the state of the river are too often neglected, with disastrous consequences to works of this kind, just on the eve o completion, or to the lands above the dam in consequence of flooding caused by the obstruction of the dam. In cases where this last danger is apprehended, a self-acting dam has sometimes been employed, consisting of a massive frame of planks carried across the river and attached by hinges to the crest of the dam. This plank is maintained in a vertical position in ordinary conditions of flow by balance weights attached or hung over wheels upon the wing walls, so as to retain the maximum desirable head of water. In floods, the increased pressure of the overflowing water overcomes the balance weights and throws down the plank into a horizontal position, opening a free passage for the water."

We have given, in preceding chapters of this work, a large number of plans for the construction of dams, applicable to the different situations and conditions which are likely to occur, and based upon general experience and the essential principles which are to be kept in view in an undertaking of this kind. The remaining chapters will be chiefly devoted to the illustration and description of dams which have actually been built or are now in process of erection in different localities in our country. This method, while it may give less scope for original suggestions and the discussion of useful theories bearing upon our subject, will give our readers an eminently practical view of the extent of some note-worthy enterprises of this class, and the minor details involved in their prosecution.

CHAPTER XXII.

DAM AT LAWRENCE, KANSAS.

The engraving herewith presented is one of a series executed by our artist from the original plans of a Dam now being built across the Kansas River at Lawrence, Kansas, and constituting a work of such magnitude and possessing such a degree of practical interest that we deem it worthy of minute description and illustration. The Kansas river is about five hundred miles in length, and drains upwards of fifty thousand square miles of land west of Lawrence. The minimum flow of water at Lawrence is about three hundred thousand cubic feet per minute, and has an average declension of about three and one-half feet to the mile. The water having a rapid flow, the river rises slowly in times of excessive rains, and reaches but moderate hights; the greatest rise within a period of four years being only six feet; and the water did not remain at that hight more than twelve hours.

The dam is intended to raise the water eight feet; though it is being built with a base sufficient to carry a dam ten feet in hight, whenever it may become necessary; and the river banks are sufficiently high to prevent overflow. The average width of the river is about six hundred feet. The length of the dam, including head-gates, piers, and abutments, is seven hundred feet. The river bed, for three-fifths of the distance across from the south bank, is solid rock bottom; the remainder, to the opposite bank, being composed of coarse sand and gravel with occasional strata of blue clay, the first of which occurs at eight feet from the surface of low water. The dam is provided with two canals, one on each side of the river. The canal on the south side is sixty

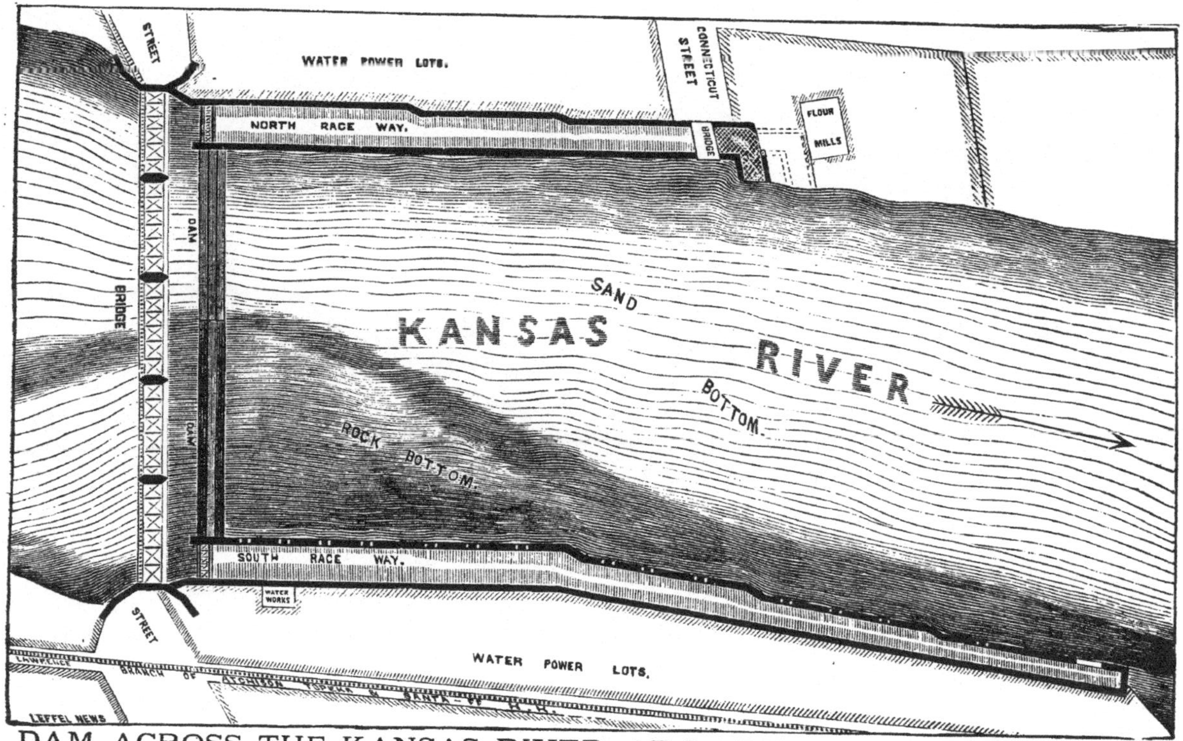

DAM ACROSS THE KANSAS RIVER, AT LAWRENCE, KANSAS.--Fig. 1.

feet in width, and that on the north side fifty feet, both to be provided with suitable head and tail gates.

Our present illustration shows only the ground plan or map of the Lawrence dam, including also a portion of the stream and the vicinity on both sides of the river. The location and relative portion of all parts of the structure are clearly shown, as are also the water-power lots, the bridge above the dam and those across the two race ways, the railroad passing the whole length of the water power tract on the south side of the river, the location of several streets on each side, and that of the Flouring Mills on the north and the city Water Works on the south bank. The Water Works are to be operated with power obtained by means of the dam, the contract with the city stipulating that for a fixed sum it shall be entitled to use sufficient power to raise two million gallons of water one hundred and fifty feet high every twenty-four hours. This is but an insignificant fraction of the aggregate power which the dam will furnish, the total amount which will be afforded when the dam is completed according to the present plans being estimated at 2,500 horse-power; and by the increase in hight which can be made, as already stated, whenever it shall be found desirable, from 1,000 to 1,500 additional horse-power will be secured.

We reserve for our next chapter the details of construction of the dam, which will be fully illustrated, showing the thoroughness and intelligent skill with which the enterprise is conducted, and the reliable character of the structure as regards the vital qualities of strength and durability. The work is carried on by Mr. Orlando Darling, of Lawrence, a civil engineer by profession, and a man of rare energy and practical judgment; and the terms upon which he is to prosecute the undertaking are embodied in a contract between himself and the city of Lawrence. We are indebted to the courtesy of Mr. Darling for material both for this article and for our illustrations. The feasibility of building a dam across the Kansas river (or the "Kaw," as it is locally and familiarly known) and obtaining thereby manufacturing facilities which would tend to greatly increase the wealth and population of the city, has been actively discussed by the people of Lawrence for nearly ten years past, or ever since the close of the war gave the town a secure prospect for the future. We are glad to know that the enterprise has at last taken a practical and tangible shape, and is in competent hands, insuring its steady prosecution and successful completion. According to our present information, (Feb., 1873,) the work of building the piers, which was begun last fall, will be completed in the coming Spring; and after the "June rise" is over, work will be begun upon the dam itself. It is intended that the whole structure shall be completed in the fall of 1874.

The city of Lawrence had an eventful early history, and has experienced many trying changes of fortune, but has bravely held its own from the first, and has for some years past been steadily advancing. We trust that the building of the dam will prove the introduction of a still more prosperous era.

CHAPTER XXIII.

DAM AT LAWRENCE, KANSAS.—(*Continued.*)

We illustrated, in connection with the last chapter, the ground plan, position and immediate surroundings of the Dam across the Kansas River at Lawrence, now in process of construction, and destined, if we may judge from appearances, to prove a very fortunate and valuable enterprise to all concerned, and especially to the people of Lawrence. Nothing so stimulates the growth of a town, and secures its lasting prosperity—unless it has a commanding commercial position such as but few cities can enjoy—as the possession and use of ample manufacturing facilities; and any community may therefore justly congratulate itself upon the creation or development of so prolific a source of wealth. As between commerce and manufactures, an inland town is usually compelled to rely upon the latter, a dozen or so seaboard, lake and river cities having a substantial monopoly of the wholesale carrying and distributing trade of the country; while manufactures flourish where neither ships nor steamboats can float, provided the power to drive machinery is afforded, and the energy to turn it to account is not wanting.

The river at Lawrence, as stated in our former chapter, has a rock bottom for a distance from the south bank of three-fifths of the width of the stream, the remaining two-fifths, on the north side, being chiefly sand and gravel. That portion of the dam which rests on rock bottom is 315 feet in length from the inside of the head-gate pier of the south gate way, and is to be built of cut stone of large dimensions, thoroughly cemented. We present in the accompanying engraving this portion of the dam, with the head gateway and canal on the south or Lawrence side of the river, showing also the city water-works, and the railroad in close proximity, as has already been indicated in the plan. The dam is represented as if cut in two a short distance from the head-gate pier, in order that its internal construction may be clearly seen. The north portion of the dam, resting on the sand and gravel bottom, does not appear in this cut, but will be fully illustrated in our next chapter.

The base of the rock dam is twenty-one feet, the whole width on top to be covered by one stone eight feet in length, with its upper corner beveled off one foot, leaving a flat surface on top of seven feet. The sides of the dam present angles of forty-five degrees with the base, and the water falls in cascades over the series of steps, any injurious effects on the foundation being thus prevented. The sides of the dam are composed of stones of not less than six feet in length and eighteen inches in thickness, laid in transversely or crosswise of the dam, the center being filled with concrete. The dam, as has already been stated, is intended to raise the water eight feet, but has sufficient base to carry a dam ten feet in hight if found desirable. The manner in which this portion of it is constructed is so clearly shown in the cut that with the aid of the brief explanation we have given it will be fully comprehended by the reader.

The head-gate way and canal on the south side, here represented, rest on solid rock, it first being excavated to below low water mark, and the water from the wheels being discharged through the arch-ways in the inside wall into the river below the dam. The floor of this canal will be hereafter described.

The stones for the masonry are all to be procured from a remarkable quarry of very durable stone, situated in Jefferson county, on the line of the Kansas Pacific Railway, ten miles west of Lawrence. From this quarry an unlimited amount of stone, in square blocks exactly eighteen inches in thickness and of any length and width desirable, may be procured. The two canals in connection with the dam are about one-half mile in length, and the Water Power Company owns and controls all the lands lying upon the canals, for their entire length. The water-power lots are from 150

DAM ACROSS THE KANSAS RIVER, AT LAWRENCE, KANSAS.--Fig. 2.

to 450 feet in length, lying in the center of Lawrence. The railroad track on the south side of the river, which is shown in the cut, is about ten feet higher than the top of the outside or bank wall of the canal, which is eighteen feet high, above low water. The inside or river wall, below the piers, is nine feet high, being only two feet higher than the dam, and will serve as an overfall for the water in time of flood.

The construction of that part of the dam resting on the sand and gravel bottom will be described and illustrated in our next chapter, together with some additional particulars in reference to the canals and head-gateways. This remaining portion of the dam comprises 210 feet of the distance across the stream, inside of the head-gate piers, having but about two-thirds the length of the part built on the rock bottom; but its construction presents more difficult problems in engineering, and more interesting features as regards the selection of materials and their employment in such a manner as to secure the requisite strength and solidity, than any part of the work thus far described. A rock bottom gives a natural foothold for the superstructure of a dam which greatly simplifies the task of the builder; but upon a soft bottom his ingenuity and fertility of resource are called largely into exercise. Our next chapter will show how in the present instance the demands of the case are to be fully met and a strong and reliable structure completed, making the whole work, what we believe it will ultimately be found, an achievement of engineering skill to which its builder and the people of Lawrence can point with satisfaction.

CHAPTER XXIV.

DAM AT LAWRENCE, KANSAS.—(*Continued.*)

We give in this chapter our third and concluding descriptive article and illustration of the dam at Lawrence, Kansas. In our first engraving the ground plan of the entire dam and both banks of the river were shown; in our second the position of the dam resting on rock bottom, extending from the south bank 315 feet or about three-fifths of the width of the stream, including also the canal on the south side, the head gateways and the water-works building; and in the illustration here given we represent the remaining part of the dam, resting on a bottom of sand and gravel, and extending 210 feet, from the north bank to the end of the rock-bottom portion, with which it connects. The canal on the north side and the head gateways are also shown; and the construction throughout is so clearly delineated in the cut that with the aid of the following description it will be readily understood, and the reader enabled to appreciate the care and thoroughness with which the work has thus far been conducted, and will be carried to completion.

The portion of the dam here illustrated is of the kind known as a frame or crib dam, and rests on a foundation of trees and rock. The bed of the stream is first reduced to a uniform depth of about five feet for a space of eighty feet up and down the stream. This being done, a tree foundation is placed in the river, composed of trees from sixty to eighty feet in length, with the limbs entire and the tops up stream. The trees are laid as close together as possible and fastened together at each end, the trunks touching each other as they rest in the foundation, and the tops interlocked so as to form a solid mat of brush. The whole constitutes a close platform, and is sunk to the bottom with small rubble stone. Five of these courses or platforms, each constructed as above

DAM ACROSS THE KANSAS RIVER, AT LAWRENCE, KANSAS.--Fig. 3

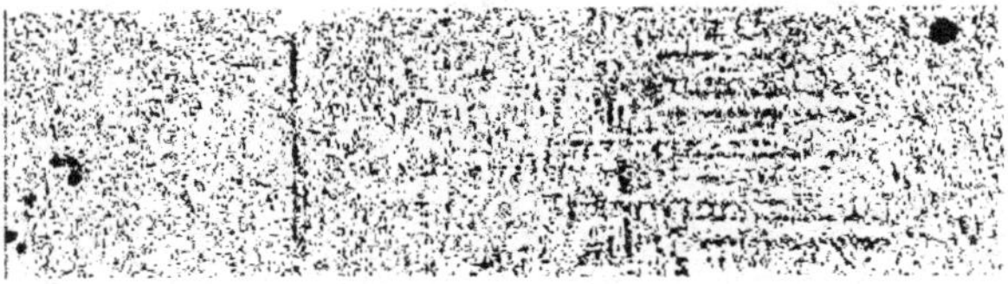

described, are laid down to complete the foundation, and each succeeding course is drawn five feet up the river, as will be seen by the appearance of the butts of the trees in the cut. As each course or platform, or part of a platform, is sunk, it is thoroughly weighted, and all interstices in the brush or between the trunks are carefully filled with rubble, so as to form a solid concrete. The frame consists of eight platforms of timber, composed of two main sills 12 by 14 inches, laid flatwise, halved together where the ends join, and bolted with drift bolts of inch iron. The main sills are held in place by means of cross-ties 8 by 8 inches, the ends of which are dovetailed into the sills. The ties are put in at intervals of eight feet, and spiked to the sills with $\frac{3}{4}$ inch drift bolts. The first platform, forming the base of the dam, is twenty-four feet wide from out to out, and is gained into the upper platform of trees so as to receive a uniform and level bearing, and each succeeding course is thoroughly secured with iron bolts to the course below, receding on each side one foot. The breasts of the dam are thus given a slope of 45 degrees, and the top is finished with a level surface eight feet in width. An open platform twelve feet in width is also firmly attached to the lower sill of the upper slope of the dam, which practically makes the base of the structure thirty-six feet in width. As each platform is laid and adjusted, it is carefully filled with rubble rock of large size, all interstices being packed with spawls so as to form a solid body.

The frame being completed and filled with rock as above described, the outer corners of the sills are trimmed off so as to present a bearing surface of four inches, and the whole is covered with plank $2\frac{1}{2}$ inches in thickness, placed in line with the current of the stream, and firmly spiked on with six-inch boat spikes. The dam having been built and completed in sections, as each section is planked over, a protection is placed on the up-stream side, extending thirty feet up the river, having a thickness of fully four feet on the open platform, and extending at least half way to the top of the dam on the upper slope. This protection is composed of large stones, all the interstices being filled with spawls.

After a row of tongued and grooved sheet piling has been driven at the lower ends of the tree foundation, a large amount of rubble rock is placed upon the lower ends of the logs and in the bed of the river below. The plan contemplates the use of 10,000 yards of rubble stone, which is being quarried out of the south bank of the river, immediately below the dam.

The head gateway and canal on the north side, which are represented in the engraving, rest on ten rows of piling driven closely together, up and down the river. The piles, after being sawed off two feet below low water mark, are capped by timbers 10 by 16 inches, laid flatwise, upon which is laid a close floor of ten-inch timbers; and this timber floor is covered with a floor of three-inch planks, laid in line with the stream and firmly spiked down with six-inch boat-spikes. On this platform the masonry is laid, each pier resting on three rows of piling driven twenty-five feet to the bed rock. Sheet piling is driven to the clay, from the bank in front of the head gates, and along the base of the inside pier to the lower end of the tree foundation, thence under the canal to the river bank, and along the bank outside of the round piling, to prevent any washing of the bank by the discharge of the water from the water wheels. Sheet piling is also driven along the lower ends of the tree foundation to the intersection of the rock bottom.

The canal below the tree foundation is constructed the same as immediately below the head-gates, excepting that the sand and gravel is washed out from among the piling to a depth of about five feet, and then riprapped thoroughly with stone to prevent any further washing. Water wheels may be located on either side of this canal by merely cutting a hole in the floor to receive the tube of the wheel, and the water is discharged under the canal into the channel of the river below the dam. The outside or bank walls below the piers are six feet thick at the base, battered on the outside to three feet on top, are eighteen feet high above low water, and will serve as a foundation for buildings their entire length, as power can be taken at any point along the canals. The piers in which the headgates are placed are six feet thick, without batter on either side, by eighteen feet in hight. The inside or river walls are six feet thick without batter; and, as already stated, are nine feet high, being only two feet higher than the dam, and thus serving in time of flood as an overfall for the water.

The head-gate piers or walls on the south side of the stream are similar in construction to those above described, except that they have a foundation of solid rock instead of resting upon piles. The floor of the south canal is also the same in general construction as that of the north canal, except as regards the foundation, the posts or piers supporting it being placed on the excavated rock bottom of the canal. The manner in which the water is discharged from the wheels has been fully illustrated, archways being provided for its escape from the south canal, while from the north canal here shown it has free passage outward both at the lower end and at the side, between the piles by which the flooring is supported.

The great work which has been thus minutely described and illustrated in three consecutive chapters is expected to reach its completion in the coming autumn (1873). We have already acknowledged our indebtedness for material for our account to Mr. Orlando Darling, the enterprising contractor, under whose immediate supervision the work is carried on, and by addressing whom at Lawrence, Kansas, any particulars which we have omitted to give, or information in regard to the leasing of power, can be obtained. Lawrence is at present a flourishing city of some 12,000 inhabitants, and has been built up to its present proportions chiefly by the excellent trade, which by its central location in a country of unsurpassed fertility, aided by its ample railroad facilities, it has been able to command. It may be for this reason that its people have hitherto been content to allow its magnificent water power to remain so long unimproved; but they have now resolved to make their town the manufacturing center of Kansas. Railroads already run out from the place in six different directions, and two more are being built into the city. The wool of California and New Mexico and the cotton from the Red River bottoms and

Northern Texas pass through Lawrence on their way to eastern market, to be manufactured and returned, lad meanwhile with numerous profits and transportation charg Nowhere, it is said, can black walnut lumber for furnit be bought cheaper than at Lawrence; and timber for ag cultural implements can be obtained at moderate pric Paper materials are abundant and cheap; yet for hundre of miles no paper-mill is found. Few points, therefore, o more attractive inducements to manufacturing enterpri than the beautiful and thriving city of Lawrence.

CHAPTER XXV.

DAM ON THE TASSOO RIVER, HINDOSTAN.

The utilization of water power for manufacturing purposes is the ordinary. in fact the almost universal object of the construction of a dam; and it has been with reference to this leading purpose that our illustrations thus far have been designed. There are, however, other useful ends for which a substantial and durable dam is sometimes requisite; and among these is the provision of the necessary supply of water for the daily use of the inhabitants of a large city. Structures of this kind and for the object indicated are to be found in various localities in our own country; but the one we have chosen for our present illustration belongs to a remote quarter of the globe and is the work of a foreign nationality; and it may possess the greater interest to many of our readers from the fact of its distance and novelty, as well as its great magnitude and the special necessity which it is to meet—that of furnishing to the people of Bombay, one of the largest cities of Hindostan, an abundant and unfailing supply of water.

The city and island of Bombay, which have nearly 800,000 inhabitants, are supplied with water from Vehar, an artificial lake in the hills of the neighboring island of Salsette. The Vehar reservoir, which was constructed by the Government of Bombay about thirteen years ago, passed into the hands of the municipality of Bombay in 1863. It is nearly sixteen miles from Bombay Cathedral and has hitherto amply supplied the wants of the island; but the rainfall of 1871 being very small, the lake, at the end of the "monsoon," (the periodical wind which blows half the year in one direction and the other half in the opposite, and which in the Indian Ocean blows from the southwest from April to October, bringing heavy rains, after which it changes to the north-east for the rest of the year) was nearly ten feet lower than usual. Attention was thus drawn to the possibility of a short rainfall in the ensuing year, 1872, from which a deficiency of water would result, with all its consequent evils. To prevent so great a calamity it was decided by the municipality to make a new lake at Toolsi, to supplement Vehar. The valley of Toolsi is 112 ft. above the top of the Vehar lake, and is divided from it by only a slight ridge of hills. Hitherto the waters flowing from the hills into the Toolsi valley have found their exit by the river Tassoo, the source of which is at the end of the valley opposite to the ridge dividing it from Vehar, whence it flows past Kennery to the sea. By damming up the source of the Tassoo, the water is impounded in the Toolsi valley, and by tunnelling through the ridge between Toolsi and Vehar, a passage is made for it into the latter lake. Of course, the supply of water into Vehar from Toolsi can be controlled, and, if not wanted, can be kept impounded in Toolsi lake till it is required, any surplus flowing over the dam across the Tassoo, and escaping by the old route. The view given in our illustration is of the dam across the Tassoo, as it now appears, 30 ft. high, with the water overflowing. This view shows but a very small portion of the intended lake, the greater part of which lies behind the low ridge stretching across the picture. It is intended to raise the dam to a height of 71 ft. The lake, which is but an auxiliary supply, has an area of 300 acres of water, containing 1,451,000,000 gallons (besides as much of the available rainfall of Toolsi as can be turned into and stored in the Vehar lake.) All this water, except a few gallons, can be made, by opening the penstock in the tunnel in the ridge dividing Toolsi from Vehar, to flow into the latter lake, and thence through the main to Bombay. Vehar, when full, covers an area of about 1,400 acres, and has 2,550 acres of gathering ground, and contains 10,650,000,000 gallons, giving a daily supply of 10 gallons a head, and Toolsi will increase it by 4½ gallons per head. This additional cheap water supply, which will probably last Bombay for twenty or thirty years, is expected

DAM FOR THE NEW WATER SUPPLY OF BOMBAY, HINDOSTAN.

to cost only four lacs of rupees, a sum equal to about $220,000 in United States currency.

The construction of the dam is not shown in detail in our engraving, but a sufficient portion of the structure is visible to give a very clear idea of the plan upon which it is built. The main feature of the work is the broad and solid wall of masonry, of which most of the lower or down stream face is shown. The up-stream face of this wall is covered and protected by strong piers of timber, between which are upright planks, closely fitted and strongly secured; the top of this part of the dam being finished with horizontal stretchers extending between the piers. Midway of the stream a space is provided for the overflow of the surplus water, which afterwards falls over the stone work of the dam into the chasm below. On the upstream side of the timber face of the dam is a filling of stone and gravel sufficient to protect the timbers and the foundation from undermining or leakage.

These extensive and admirably constructed works were designed and carried out by Mr. Rienzi Walton, an associate member of the Institute of Civil Engineers, who is the acting executive engineer of the municipality of Bombay.

CHAPTER XXVI.

LOCK AND DAM AT HENRY, ILL.

We have described in preceding chapters the construction of dams for the utilization of water power, for the measurement of streams, and for the provision of a supply of water for the use of the inhabitants of a populous city. We now give illustrations and a general descriptive sketch of a dam, the object of which is to improve the navigation of one of our Western rivers and afford, eventually, a channel of traffic between the lakes and the Gulf of Mexico, which will be of immense value to the people of our great grain growing States. The most urgent need of the agricultural population of the Mississippi and Ohio Valleys is cheap transportation for their products, by which the value of their wheat and corn shall not be wholly absorbed by the charges paid in getting it to a good market. As the case stands, the railroads do not afford this indispensable means of conveyance at a reasonable cost; and without discussing the question where the fault, if any exists, is to be found, we may safely say that no greater blessing could be bestowed upon the people of the West than direct communication by water with the great centres of commerce.

With a view to securing this most desirable end, the State of Illinois has undertaken to improve the Illinois river so as to make it navigable for the largest class of steamers that traverse the Mississippi river at all navigable seasons of the year, from the mouth of the Illinois up to Lasalle, the point where the Illinois and Michigan canal enters, 100 miles southwest from Chicago. The distance thus comprised is 230 miles. The improvement is to be made by the construction of locks and dams, forming a slack water navigation. The locks are to be 350 feet long between the gates, and 75 feet wide, and the dams are to be raised so as to make a

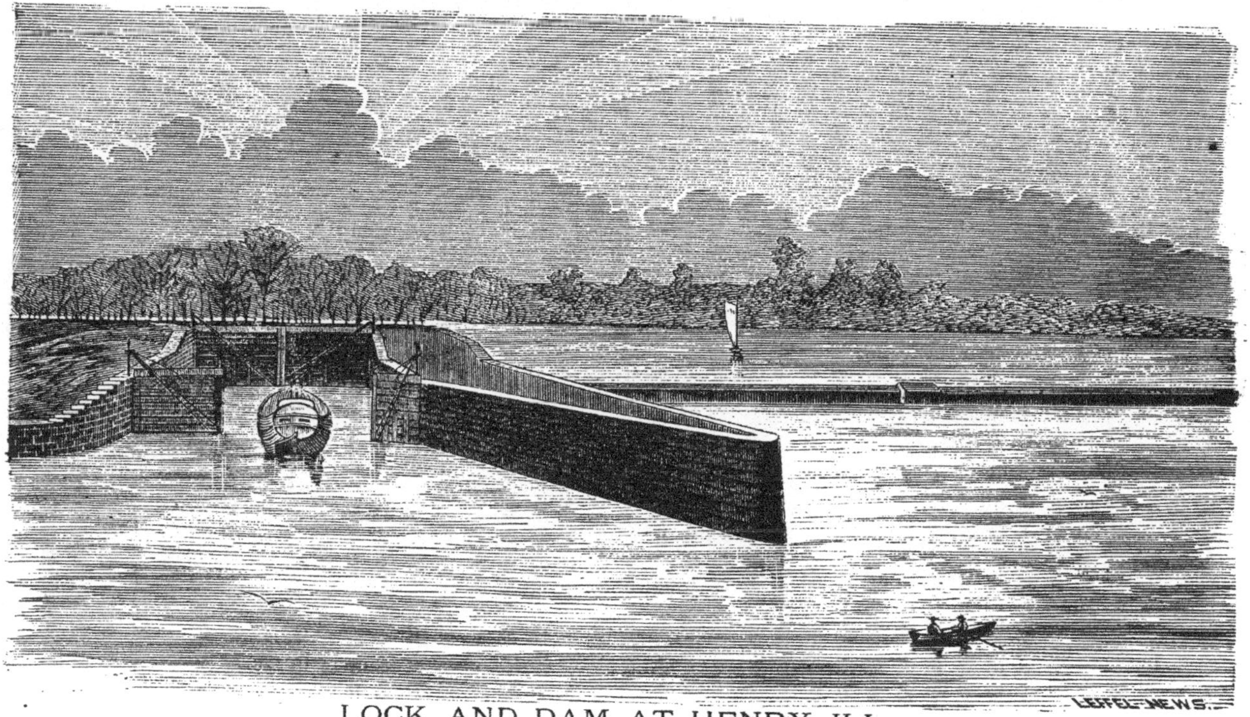

LOCK AND DAM AT HENRY, ILL.

uniform depth of 7 feet of water for the entire distance at all seasons. At present, during low water, there is less than three feet of water on many of the bars in the river, and on one only 16 inches.

For the entire improvement, from Lasalle to the mouth of the Illinois river, five locks and dams will be required, the total cost of which is estimated at $2,200,000, being less than $10,000 per mile for the 230 miles comprised in the work; and this certainly appears a moderate expenditure for the facilities of transportation which will be secured. One lock and dam was finished some time since, at a cost of $400,000. This is the work illustrated in our engravings, its location being at the town of Henry, 28 miles below Lasalle and 32 above Peoria, the second largest city in the State. The lock is on the north side of the centre of the river. The dam connects with an outside protection wall, about 100 feet above the upper gates, and a short distance above the bridge erected not long since at this point; and joins the south bank midway between the bridge and the mouth of Sandy Creek.

The two smaller engravings of the three which we here present, give front and sectional views of the dam, the top of one end of which may also be seen at the right of the larger cut. The dam is built of timber cribs filled with loose stones. It is 35 feet wide, 11 feet high, and 540 feet long, and is raised 6 feet above low-water mark. On the upper side of the crib-work sheet-piling is driven into the bed of the river to a depth of about 5 feet, and on the lower side piles 12 inches square are driven close together, 10 feet into the bottom of the river. Twenty feet of the width of the dam has a coping of timber, sloping up-stream, 8 inches thick at one end and 4 inches at the other. There are two drops of 3 feet each on the lower side, with solid timber aprons twelve inches thick, and 7½ and 8½ feet width on which the water falls. Below the lower apron and the square piles, stone and brush are extended 20 feet on the river bed, and above the dam for 50 feet it is filled in with brush and gravel, tapering over on to the upper slope of the timber coping. The dam is securely bolted at every crossing of timber, and there were used in its construction more than 20,000 lbs. of wrought-iron bolts, generally ¾ inches square and 14 to 22 inches long.

In our illustrations, Fig. 2 gives a sectional view of the dam, showing distinctly all the parts as above described. In Fig. 3 is given a front view of the dam, such as would be obtained by standing midway in the stream below the dam and looking up-stream. By comparison of the two figures, the different parts will be readily identified.

The lock is built entirely in the bed of the river. In commencing the work, an area of a little over seven acres was

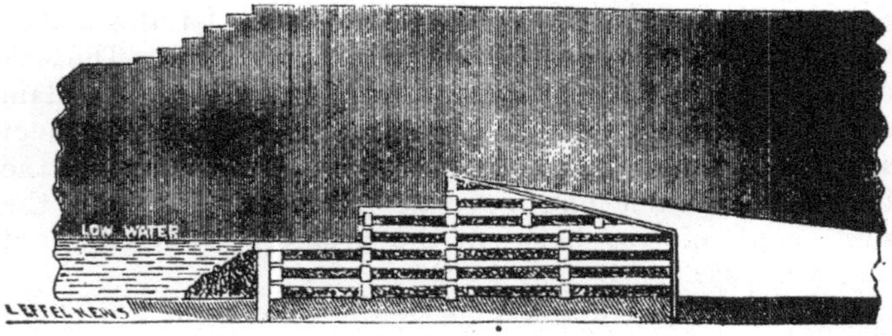

FIG. 2.

inclosed by a coffer-dam substantially built of piles with cap timbers, and of sheet piling driven outside from 6 to 10 feet into the bed of the river, and protected with gravel on the outside of the piling from the bottom of the river, with a suitable outside slope. The water, which averaged only 4 feet in depth over the whole area at low water, was removed by a rotary clam-shell pump, with an iron delivery pipe 9 inches in diameter, and driven by a 15 horse-power steam-engine. The area of pit to be excavated for the foundation of the lock was 485 by 115 feet, and averaged 6 feet deep, requiring the removal of 13,000 cubic yards of earth. After the excavation was made, 3,200 bearing piles of hard wood, from 12 to 25 feet long and 12 inches in diameter at the large end, were driven over the bottom. On these piles eleven rows of square timbers, 12 by 12 inches, were placed longitudinally, each row extending 477 feet, and secured to the piles with bolts 22 inches long and $\frac{3}{4}$ in. square. On these timbers cross timbers 12 inches square were placed 6 inches apart, covering two thirds of the whole area, and bolted to the piles and the longitudinal timbers at every crossing with bolts $\frac{3}{4}$ in. square. All the spaces from the top of the cross timbers to 3 inches below the bottom of the longitudinal timbers—a depth of 27 inches—were filled with concrete. The whole of this foundation was covered with $2\frac{1}{2}$ inch planking secured to the timbers.

On this substantial foundation the side walls of the lock were commenced, extending 476 feet on each side, with a mitre sill wall under the upper gates, and a breast wall at the head uniting with both side walls, which are 73 feet apart at foundations through the lock chamber. For 176 feet from the head of the lock the walls are 30 feet high, and for the remaining 300 feet 24 feet high; the upper end of the lock being 6 feet higher than the lower end, on account of the extreme depth of the water in time of floods. The main walls of the lock, where the hight is 30 feet, are $11\frac{3}{4}$ feet thick, while those 24 feet high are $10\frac{3}{4}$ feet thick at the foundation. The breast wall is from 7 to 8 feet thick and 7 feet, 8 inches high. The mitre sill wall is of the same hight, and from 10 to 15 feet thick; and in this wall are eight arched culverts $5\frac{1}{2}$ feet wide and $3\frac{1}{4}$ feet high, through which water is admitted into the lock. Below the lower gates the main walls extend 20 feet, with wing walls 40 feet long on both sides, flaring 10 feet each at the lower end. The water is discharged from the lock through semi-circular arched culverts 5 feet wide, with abutments 6 feet high, and connected at the top with an arch of $2\frac{1}{2}$ feet radius. The masonry is composed of magnesium limestone and is very substantially put up, laid in the best quality of hydraulic cement mortar; and amounts in all to 10,328 cubic yards. The entrance to the lock is formed by heavy rubble walls extending up from the head of the lock on each side. On the north or shore side it turns with a curve towards the shore, and joins a slope wall protection to a guard bank, which extends 350 feet to connect the lock bank with the shore. On the south, or river side, this wall extends 100 feet, flaring out from the line of lock walls; then, forming a circular pier-head of 8 feet radius, it extends down parallel with the lock walls (the faces of the two walls being 50 feet apart at bottom) to 100 feet below the end of the lock, where it forms a pier-head similar to the one at the upper end, and returns to join the wing wall of the lock. The entire length of this wall is 900 feet on the river side, forming the abutment to the dam and the river protection to the lock. It is from 20 to 29 feet high, 7 to 8 feet thick at bottom, and 3 feet thick on the top, with a stone coping 9 inches thick. At the foot of the lock, on the shore side, is a similar wall extending down 50 feet and then curving towards the shore. This

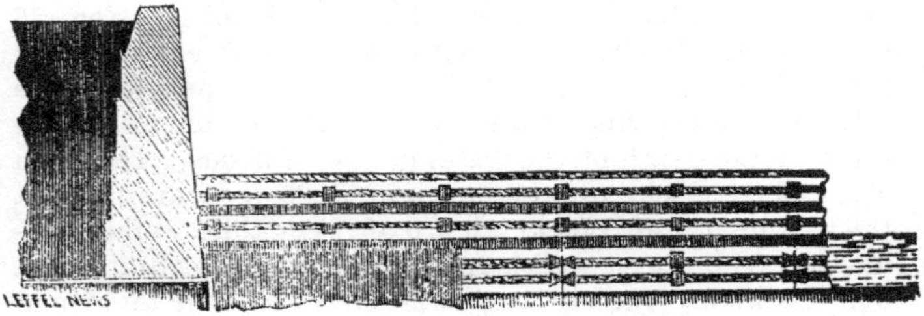

FIG. 3.

wall is but 15 feet high. The rubble walls are all on a foundation of piles, timber and planking, there being 860 piles, from 12 to 19 feet long. There are 5,560 cubic yards of rubble wall, of which 5,300 cubic yards are laid in hydraulic cement, and 250 cubic yards are laid dry.

The filling between the outermost or river wall and the wall of the lock with its wing extension, is not shown in our engraving, the space being represented as if hollow, thus showing the interior or bank side of the wall, as it appeared before the filling was put in.

The lock-gates are of massive proportions, 24 feet high and 43 feet wide, each gate containing over 20,000 feet, board measure, of the best white oak timber, and 27,000 lbs. of wrought and cast iron (including the anchor irons by which the gate is secured to the lock walls) weighing over

60 tons, and costing, with the hanging fixtures, $4,000 each. The mechanism for operating the gates and for admitting and discharging the water is all of the strongest and most complete description, and the balance of the gates is so perfect that, ponderous as they are, two men can open or close them in four minutes with ease. The lock can be filled or discharged in three minutes, moving at its maximum lift 172,000 cubic feet of water. A single boat can be locked through in about fifteen minutes, a fleet requiring more time, as the boats have to be got into position in the lock. The capacity of the lock is equal to 12 canal boats at one time, of the size of those on the Erie or the Illinois and Michigan Canal.

This extensive work was completed Jan. 11, 1872. It was prosecuted under the official charge of Messrs. Joseph Utley, Virgil Hickax and Robert Milne, canal commissioners, the designing and construction being the work of Daniel C. Jenne, chief engineer, assisted by Geo. A. Keefer, John S. Butler and Chas. C. Upham; and the details were faithfully carried out by the contractor, William Johnson. The stone was obtained from the Semont and Joliet magnesium limestone quarries, the oak and pine timber from Michigan, and the other timber and piles from the immediate vicinity of the work.

By the building of this lock and dam, 60 miles of good navigation is added to the Illinois and Michigan Canal. It is intended eventually to extend this work to Chicago, by improving the river in a similar manner 60 miles to Joliet, and enlarging the Illinois & Michigan canal to the Chicago river for 36 miles. It is estimated that this will cost over $16,000,000, but the money will be well invested in an enterprise so vastly beneficial to the people of the west and southwest, giving to them, as it will, direct communication from Chicago by large steamers to New Orleans and to all points on the Mississippi river and its chief tributaries. A total of 1,500 miles of navigation will thus be secured, and a cheap channel of commerce opened for an almost boundless agricultural region, which needs for its complete development only the means of transporting its products at small cost to the distant points where a remunerative awaits them.

CHAPTER XXVII.

CRIB DAM WITH PLANK COVERING.

Crib-work, when properly constructed, with suitable filling and secure fastening of the timbers where they cross, is as reliable an arrangement for a dam as any that can be mentioned. It of course requires the use of a considerable amount of timber, there being, from the nature of the structure, no chance to economize in this respect; but as crib-work can be put up in a thorough manner with the employment of but very little skilled labor, the saving in this point is often much greater than the reduction in cost of material would be by adopting any other plan. Moreover, even in a country not heavily timbered, there is apt to be in the vicinity of a water-course a sufficient amount of such timber as will answer for crib-work, with a little management; and the builder need not therefore, in most cases, resort to an unsatisfactory style of construction, or pay high wages to expert professional workmen, unless an unusual scarcity of timber prevents him from adopting such a method as is described in the present chapter.

The plan of dam here illustrated is suited to comparatively narrow streams, not exceeding 100 feet in width, and where not over 10 feet head of water is to be afforded. The material used is timber, brush and loose rock, with an admixture of fine or coarse gravel or coarse sand and dirt. The crib is a continuous one, from bank to bank, the timbers crossing the stream being spliced together where required. At intervals of 8 or 10 feet timbers are placed crosswise of the dam, in the direction of the stream, their purpose being to bind the frame together. They are firmly secured to the longitudinal timbers with either pins or spikes. They diminish in length from the base to the top of the dam, as the up-stream and down-stream sides of the dam have each a slope which brings them nearly together at the top. The filling may be of brush and clay, though gravel is preferable to clay under all ordinary circumstances. In this case, as the dam is intended to be as nearly water tight as may be, the clay will serve very well to give weight and solidity to the crib-work. For the purpose of excluding the water, the dam is covered tightly with plank on both sides and on the top; and an apron of planks, supported on logs, is also placed in front of the dam down stream.

For abutments, square cribs are built, of the same general nature as the dam, but of somewhat greater width and hight. The cribs should be filled with stone and gravel, and thus made as heavy and substantial as possible. It is an excellent plan, also, to give the dam a curve up stream, as shown in several preceding chapters, its strength being thus considerably increased.

The dam in our engraving is supposed to rest on a hard bottom; but should the bed of the river be soft, it would only be necessary to put down in the first place a foundation of logs, laid close together, lengthwise of the stream, and projecting beyond the base of the dam down-stream.

Should leaks occur in the plank covering of the dam, it may be made tight by stirring saw-dust or fine tanbark into the water above the dam. A small leak at the bottom may be stopped by crowding straw or fine brush into the hole and covering it with earth. The exact locality of such a leak may be found by stirring a little saw-dust into the water near the bed of the stream and observing where the current carries it through. For a larger leak, the best plan is to put in first a stone nearly large enough to fill the hole, following it with smaller stones and finally gravel and loam. This has the advantage of being more durable than the straw filling, which will rot in time.

Although relating to a different style of construction from that above described, we may venture to add in this connection the following letter on the subject here treated of, received some time since by the publishers of this work:

EDITOR LEFFEL NEWS: My model of a dam is a common

barn roof placed across the stream; the eaves—the upper one especially—sunk to a solid foundation, the peak or ridge extending level from bank to bank across the creek. Or the half of the roof down the creek may be omitted, and let the surplus water drop perpendicular on an apron. (I adopt the roof as an illustration as being so easily understood, the bearers being in place of the rafters, and the planking in place of the roof boarding.) The bearers should be supported by plates running across the stream, except at the foot, where they should be beveled and rest on the bed of the creek. The plates may rest on stone walls, log cribs, or posts standing on mud-sills that have been well bedded and in the direction of the stream. The strength of timber and distance apart must be in proportion to the height of the dam. There are dams near here in rapid streams, the foundations of which have been in for over one hundred years. WM. C. CRAWFORD.

MILFORD, PIKE CO., PENN.

We think a sloping face on the downstream side of the dam, with an apron to carry the water fairly away, is preferable to allowing the water to drop perpendicularly on the apron. It is at the foot of the dam, in front, that undermining, wearing and washing out very frequently occur, by eddies and reacting currents of water. If the water comes down in a sloping directiion to the apron, as in the dam which we illustrate, or as in the "roof" construction suggested by our correspondent, the danger referred to is almost entirely removed.

CHAPTER XXVIII.

PLANK DAM AT GILBOA, OHIO.

In the dam which we illustrate in this chapter, a combination is presented of the plank construction with stone and dirt, or gravel, for the interior filling, the latter giving the dam its requisite weight and solidity. In the Plank Dam illustrated in Chapter XVI, only a single tier of plank was laid, the main reliance for sustaining the pressure of the water being the curve of the dam up-stream, and no filling, even, being required, except a little on the upper side of the dam to keep it tight. The dam thus built, however, was suitable only for a rock bottom. A nearer approach to the construction here shown was that of the dam described in Chapter XII, in which two tiers of plank are erected; but the maner of laying them is materially different from that here represented, as in this instance the wide base and narrow top, or the pyramid construction, so to speak, add very greatly to the firmness of the structure, rendering it hardly possible for any direct pressure to force it from its foundation.

Our engraving does not represent a merely theoretical style of construction, but is a view of a dam in process of erection for supplying the power to a Leffel wheel running the grist and saw mill of A. D. McClure, Esq., located on Blanchard River, in the town of Gilboa, Putnam Co., Ohio. The stream at this point is 250 feet wide and has a rock bottom. The perpendicular hight of the dam is six feet. It is shown in the cut as if cut across in the direction of the stream, showing the ends of each tier of plank, the cross-ties connecting them, and the filling between the tiers. The planks used in the construction are 10 inches wide, 2¼ inches thick, and of any convenient length; care being taken in laying them up to break joints. The cross-ties extending from one tier to the other are planks of the same

PLANK DAM AT GILBOA, OHIO.

thickness as those in the two walls, but may vary in width, while their length is of course determined by the distance between the tiers. They are put in at every 8 or 10 feet, or thereabout, of the length of the dam, their ends coming out between the ends of two planks which meet in the tier, and flush with the outer face of the dam. Their thickness being the same as that of the other planks, a close joint is hus made and the dam remains tight.

The planks in each tier are so laid as to fall back two inches at each successive course or layer, a continuous slope thus being given to the face of the dam, at an angle of about forty-five degrees. This falling back does not begin, however, on the up-stream side, until about half the hight of the dam has been reached, the first three feet of this tier being perpendicular. The slope then commences and continues at the same angle as the down-stream face, each layer of planks falling back two inches, until the crest is reached, where the top layers of each tier are placed side by side, as shown in the cut. It may be well to surmount these with a single plank as a cap or cover, so that there will be no open joint along the crest of the dam. Built in this manner, the entire width of the base of the dam, from out to out, will be about ten feet, of which six feet or more will be on the down-stream side of a point directly under the crest of the dam, the continuous slope being on that side.

For fastening the planks either pins or spikes may be used, but pins are preferable, as, being of wood, they admit of the planks being readily sawed out when repairs are required. The planks in the down-stream tier are of oak, while in the up-stream tier sycamore and elm are used below water, and oak above. Sheeting, or even round poles, may be put on the top of the dam on the up-stream side, if desired, to guard it from ice and driftwood. The whole upper side of the dam, indeed, may be covered with cement to make it perfectly tight. Mr. McClure, who is erecting the dam here described, proposes to build without putting on either the cement or the sheeting, but will add them afterward if found expedient. Still another additional means of protecting the up-stream tier of the dam is to build a wall directly against and of the same hight as the perpendicular portion of that tier, which in the dam under consideration is the first three feet of the tier. This may be done if stone is abundant and easily obtained in the locality where the dam is built, but the necessity for it is not sufficiently urgent to warrant any heavy outlay, such as would be required if the stone were to be brought a considerable distance.

The dam built by Mr. McClure is abutted at one end against the stone wall of the mill; the other end extending into a square crib about ten feet high and built of the same material—2 or 2¼ inch planks—and in the same manner in all essential respects as the dam itself. It is about twelve feet square, and filled with stone. The crib in our engraving is shown as projecting considerably beyond the base of the dam, as we regard it very desirable that a crib should extend two or three feet beyond the dam on both the up-stream and down-stream sides. If the dam, therefore, has a width of ten feet at the base, as in the case here described, we would give the crib a width of at least fourteen feet, thus allowing a projection of two feet beyond the dam in both directions.

Care should be taken to make the dam as tight as possible on both sides, avoiding any cracks or gaping joints in the tiers of plank or at the points where the ends of the cross-ties occur. The filling between the tiers, in Mr. McClure's dam, is stone and dirt. Gravel, either fine or coarse, may be used instead; or stones of irregular size, from boulders down to small cobble-stones, mixed with loam and a moderate amount of clay—the more stone and the less clay, the better. We have shown in our engraving a filling of stones, earth and brush against the up-stream side of the dam. If sheeting, cement, or a breast work of stone is put on this side of the dam as above suggested, the filling will not require to be so heavy or so carefully put in as would be necessary if the dam were in no other way protected.

The construction thus fully described is intended for streams having a rock bottom. If the bottom is soft, a foundation must first be laid, for which purpose logs placed close together, side by side, lengthwise of the stream, constitute the most reliable material. They should be long enough to project beyond the base of the dam both up and down the stream, but especially down-stream, in which direction they should extend far enough to form an apron to the dam ; and this apron may be covered with plank to render it still more secure against the undermining action of the water. It will be observed, however, that the continuous slope of the down-stream face of the dam has a very advantageous effect in preventing the reacting, eddying and undermining tendency of the water, thus making it much less destructive to the apron and foundation of the dam. The filling or other protection for the up-stream side and upper slope of the dam would be substantially the same for this as for the dam on rock bottom.

CHAPTER XXIX.

FRAME DAM AT CLIFTON, OHIO.

A description of a dam actually erected and in use is of more value than any theoretical plan of construction can well be, not only because it has been proved by greater or less length of service to be adapted to its purpose, but also because the difficulties to be overcome, and the demands of the situation in every particular, are realized and can be pointed out, whereas in a proposed method which has never been verified in practice, many of these important details are liable to be overlooked. We illustrate in this chapter a dam which has been for some time in use, located on the Little Miami river, near the village of Clifton, in Greene County, Ohio. The course of the Little Miami in the region of Clifton, as many tourists from distant States are well aware, is remarkable for the picturesque scenery which it presents. The towering cliffs, deep gorges and shadowy ravines which have made this locality a favorite resort of pleasure-seekers for many years, fall short of actual grandeur and sublimity only when compared with the bolder natural features of the Eastern or far Western portions of the continent. In the absence of mountain ranges, which are necessary to the display of such hights and depths as are sufficient to strike a travelled observer with awe, this charming spot is as well entitled to be called a Switzerland in miniature as any thing which Ohio or the prairie States can boast.

At the point where the dam here illustrated is built, the stream has a much wider bed than immediately above or below, so that the length of the dam is about 100 feet. The bottom is solid rock, and the cliffs on either side rise to a hight of 70 or 80 feet. A rock against which the dam abuts on the right bank is itself nearly 40 feet high.

The foundation of the dam consists of six sills, 10 by 14

FRAME DAM AT CLIFTON, OHIO.

inches, crossing the stream, and placed from 6 to 8 feet apart, being somewhat nearer together under the apron than beneath the main structure of the dam. The ends of the sills are mortised into the rock at each bank, the sill farthest up-stream being also imbedded in the rock for its entire length. Across these sills are laid timbers 10 by 14 inches and about 40 feet long, the distance between centers being about 6 feet. They are secured to the lower or foundation sills by 1¼ inch barbed iron bolts, 2 feet long. Both these courses or layers of logs are squared on the top and bottom. The two center sills running lengthwise of the stream, and situated at the angle of the dam, are bolted together.

The breast of the dam is raised at the fourth foundation sill, counting from the up stream side. The upright posts constituting the front of the dam are 12 inches square, about 15 feet long, and have an inclination from perpendicular of nearly 3 feet up-stream. The posts at the center or angle of the dam, like the two upper sills at that point, are set close together and bolted to each other. The whole number of posts is the same as the number of upper sills, and they are placed the same distance apart, each post resting on a sill, into which its lower end is mortised. Upon the top of the posts, crossing the stream, is a cap timber 12 by 14 inches. This cap timber is in four pieces, each 25 feet long, spliced where they connect, and mortised into the rock at each bank. The upper ends of the upright posts are mortised into the cap-timber.

The rafters composing the up-stream slope of the dam are 10 inches square and some 25 feet long, and equal in number to the posts—each rafter, with its corresponding post and the interior braces, constituting a "bent." The rafters are mortised into the cap-timber, and between them are short cross-timbers or ties, 10 inches square, mortised into the sides of the rafters, and flush with their upper surface. The same description of ties are placed between the upright posts. The purpose of these ties is to give firmness to the frame, and prevent any tendency of the bents to sway or spread to or from each other. There are three of these ties between each pair of rafters and two between posts, so that each bent is connected with the next by five ties, the distance between the ties being from 6 to 8 feet. The foot of each rafter rests upon one of the upper sills, and a bolt is driven through the rafter and upper sill to the foundation sill of the dam.

This slope of the dam is covered with two coats or courses of 2-inch oak planks, jointed in the usual manner, the first course being laid crosswise of the stream, and the second or top course lengthwise, in the same direction as the rafters. The planks are secured by spikes, and the top course footed at the lower ends of the planks against a sill running across the stream the whole length of the dam.

In the interior of the bents are braces 10 inches square,

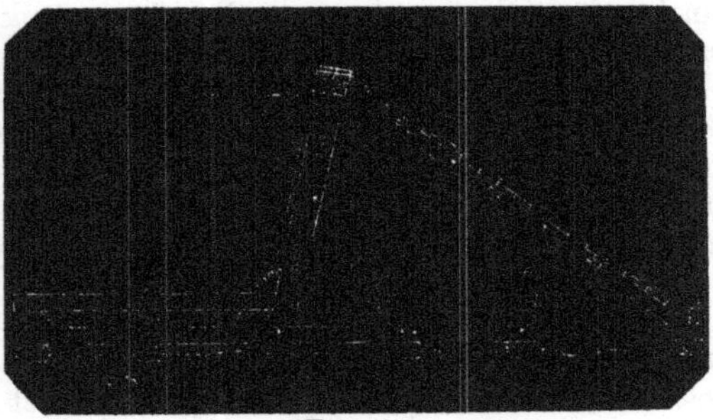

FIG. 2.

mortised into the posts, sills and rafters, in the manner and position indicated in Fig. 2 of our engravings.

The apron is constructed by placing transverse or cross sills upon the longitudinal or upper sills of the dam, these cross sills being three in number and about 6 feet apart. Upon these sills are laid timbers of the same size, lengthwise of the stream, hewed on two sides and laid close together. These logs, forming the apron, are 17 feet long. At their upper ends, where they meet the breast of the dam, a cross-timber is laid on, running the whole length of the dam, and beveled on its front so as to leave its top only 2 inches wide. The cap-timber at the top of the posts also projects 2 inches, and the 2-inch planks constituting the face of the dam, which are spiked in an upright position upon the posts and cross-ties, make with the 2-inch projections at their top and bottom, against which they rest, a continuous smooth face to the dam, giving the water an unbroken and even fall.

All the sills, rafters and ties composing the frame work of the dam are of white oak, secured by iron bolts.

The interior of the crib-work composing the apron is filled compactly with stone. The interior of the dam is also filled with stone, about half-way to the top; and against the up-stream slope of the dam is a filling of gravel and clay, extending from some fifteen feet above the dam to a point half-way up the slope, thus covering the entire lower part of the planking.

The ground plan of the dam is such that it makes an angle with the apex up-stream, instead of the arch or curve often used. The angle is but a moderate one, the center of the dam being but about 4 feet farther up-stream than the ends, while the entire length, as already stated, is about 100 feet. As will be seen by the dimensions already given, the base of the dam, inclusive of the apron, has a total length of about 40 feet.

Near the center of the dam is a waste-way regulated by a sliding gate on the upper side of the dam. The door of the waste-way is indicated in the cut near the base of the dam, close to the center posts.

The hight of the dam from the apron up to the crest is about 14 feet, the 15 feet posts having a slight inclination, as already stated. The hight from the rock bottom to the top of the dam is 17 feet or thereabouts, the top of the apron being 3 feet or more above the rock.

The water enters the forebay behind the large rock on the right bank of the stream, and only a portion of the forebay, consequently, is shown in our engraving. The race passes through an opening in the rocks, and the frame-work of the head-gates is mortised at both ends into the solid rock.

This dam was built nearly three years since by Messrs. King & Hagar, at that time proprietors of the Clifton Paper Mills, now carried on by Col. David King. A general view of the dam is given in our principal engraving, and in Fig. 2 is represented a cross-section, showing the upright post, rafter and braces constituting a bent, the two courses of plank on the up-stream slope, the end of the timber against which the upper course is footed, the upright planking on the face of the dam, the ends of the cap-timber above and the beveled timber below, and also the ends or sides of the various sills composing the foundation and the crib-work of the apron.

As the region in which this dam is situated is visited every year by large numbers of tourists, many of our readers may already have examined the structure above described, or may have opportunities of doing so in future. At a favorable season of the year, a more attractive spot can hardly be found, though there are many of wider reputation and more favored by the patronage of the wealthy and fashionable.

CHAPTER XXX.

TIMBER DAM AT NEW HARTFORD, CONNECTICUT.

The durability of the dam which we illustrate in this chapter has been proved by a period of service dating back farther than the birth of many of our readers. It was built in 1847, and has therefore stood over 26 years, requiring in all that time but little repair or alteration. It belongs to the Greenwoods Company, of New Hartford, Conn., and extends across the Farmington river at that place.

The dam, as will be seen by our engraving, is built of timber, no other material being used except that required for filling or staying the structure, and for the abutments as will be hereafter described. The timbers used are from 9 to 12 inches in thickness, the first or foundation tier being laid crosswise, the second tier lengthwise of the stream; and this arrangement is continued throughout, the alternate layers crossing each other until the work is brought to the desired hight. When complete, it has the form of a pyramid, the sides presenting an angle of 27 degrees with the horizontal line or bed of the stream. This angle of the sides gives the base such ample width in proportion to the hight that, taking this in connection with the pyramidal form, the dam has a degree of solidity and strength in its very shape which ensures its durability.

The timbers are fastened, at each point where they cross, with a spike of $\frac{3}{4}$-inch round iron, 20 inches long. The water side is covered with planking of 3-inch oak and chestnut, jointed, and put on with 7-inch cut spikes. The timbers running lengthwise of the stream are placed 6 feet from center to center, the ends coming out flush with the face and back of the dam. The timbers running crosswise of the stream are so placed as to give from two to three feet in the clear; and all the spaces are filled with stones, from the foundation up to the cap-log.

On the lower or down-stream side of the apron, piles are driven and securely fastened to the lower mud sill, on which the apron partially rests. The apron is composed of timbers 12 inches thick, placed close together. In order to attach the apron firmly to the main structure of the dam, the following plan is adopted: once in every six feet of the apron a timber longer than the others is put in, extending up-stream under the dam a distance of 25 or 30 feet, while the other apron-timbers run only 2 or 3 feet under the first timber of the dam proper. By this means, without putting more timber into the apron than is absolutely required, it is nevertheless held so firmly to the main structure of the dam that no danger of separation exists.

The entire length of the "rollway" is 232 feet. The hight of the dam, from the bottom of the mud-sills to the top of the cap-log, is 21 feet. The width at the bottom, from the foot of the dam to the up-stream side, is 68 feet; and the apron projects 14 feet beyond the foot of the dam.

The river bottom at this point consists of cobble-stones, gravel and quicksand. The banks are gravel and sand. The gravel is of the kind known as "washed," it being devoid of all the fine admixture which renders a bank tight against water. On the upper side of the dam is a filling of gravel to within 4 or 5 feet of the cap-log. It is not uncommon for the stream to rise to such a hight as to give six feet of water on the cap-log; and a depth of even 10 feet at that point has been known, but is of rare occurrence. The capacity of the river at this place is stated at 14,525,000 cubic feet of water every 24 hours during an ordinary drouth.

Our present illustration gives a perspective view of the exterior of the dam, showing also one of the abutments, which are of solid masonry, and pyramid-shaped, like the dam itself. In our next chapter we shall give a sectional view, showing more clearly the interior structure of the dam, with some additional particulars required to complete the description.

TIMBER DAM AT NEW HARTFORD, CONN.

CHAPTER XXXI.

TIMBER DAM (*Continued.*)

In our last chapter we gave a perspective view and general description of the dam of the Greenwoods Company at New Hartford, Conn., comprising the dimensions of the dam, the material used, size of timbers, and the manner in which they are put together and secured. We now present a sectional view of the same dam, from which the arrangement of the timbers will be still more clearly perceived. In this engraving, also, a full view is given of several portions not shown in the perspective cut, such as the apron in front, composed of twelve-inch timbers placed close together; the piles driven on the down-stream side of the apron, and fastened to the lower mud-sill, extending into the bed of the stream to a depth of 15 feet; and the form of the abutment, the face or front portion of which also rests upon piles. Our present engraving shows but a small part of the dam, the entire length of which, as already stated, is 232 feet. In the view it is represented as if cut transversely, in the direction of the stream, showing the internal framework, but not the filling of stones in the interior, or that of gravel in the upper side of the dam.

The strength and stability afforded by the pyramidal shape of the dam will be readily seen in this illustration, the only real source of danger being from the washing out of the gravel, especially on the lower side of the dam, which is liable to occur at a time of very high water. This difficulty did in fact present itself in the case of the dam here described, during a very heavy flood some fifteen years ago. The water acted with such effect at the lower side of the apron that a considerable quantity of gravel was washed away; to remedy which the proprietors built cribs of poles and logs, and filled them with large rocks, weighing two to three tons. These cribs were then sunk to the bottom, and the whole chained to the piles at the foot of the apron; since which time no trouble from washing out has been experienced.

About three years ago, some repairs of the dam were found necessary, and nine or ten feet of the top was taken off, the timbers having become rotten. The cause of the rotting was ascertained to be that the dam, when originally built, was planked tight on the lower side as well as on the water side, leaving no avenue of escape for the hot steam which gathered inside the dam in hot weather. The face of the dam being to the south, the heat of the sun had a powerful effect in generating this steam in the interior, with the injurious results to the timbers above indicated. All the planking on the lower side was therefore removed, and for the last three years, consequently, this side has been in the condition shown in our engraving in the last chapter.

It is proposed to raise the dam six feet in the course of next summer, making with the present hight of 21 feet a total hight from bottom of mud-sill to top of cap-log, of 27 feet. It has already been mentioned that the stream frequently rises to such an extent as to give six feet of water, and in rare instances even ten feet, on the cap-log of the dam as it now stands. The increase of hight will therefore afford the means of a corresponding addition to the amount of power held in store, the present structure being hardly in due proportion to the capacity of the stream. Our engraving shows the wall as already raised. The dam or "rollway" will be raised by placing at every 6 feet of the length of the dam a frame or trestle resting, as it were, astride of the crest of the dam and very firmly secured on both the upper and lower slopes of the present structure. The water side of this additional framework will be covered tightly with 3-inch jointed plank. Upon the lower side will be placed 3-inch planks, 2 inches apart, the object of this arrangement being to ventilate the interior and give free escape to the hot steam generated as already described.

It is believed by the proprietors of this dam that it will

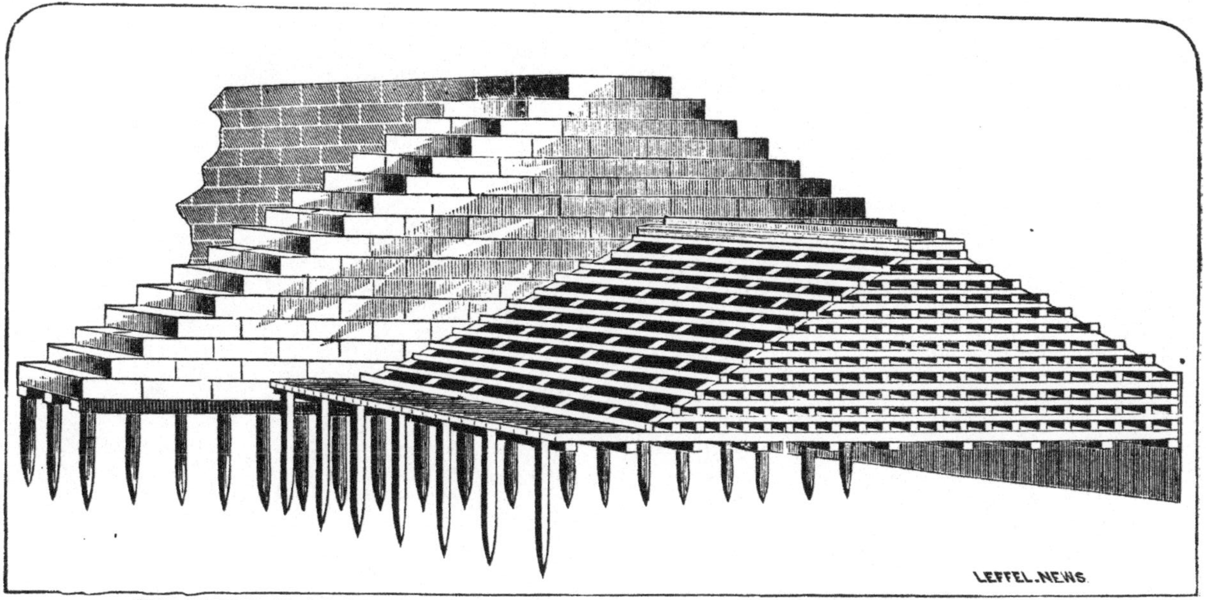

TIMBER DAM AT NEW HARTFORD, CONN.--Sectional View.

compare favorably with any other structure of the kind in the country as regards the quantity of water held in reserve by it, for the amount of money invested. It certainly possesses great merit in the plan of construction, and has been found to possess, in practice as well as theory, the prime essentials of a good dam,—strength and durability.

CHAPTER XXXII.

LOG DAM FOR NARROW STREAMS.

The description and illustration which we present in this chapter were elicited by an inquiry on the subject of Mill Dams published in Leffel's Illustrated Mechanical News for February, 1873. In order to present more clearly the suggestions embodied in the ensuing article, we first reprint in full from the Mechanical News the inquiry alluded to, (and also the comments editorially made upon it) as follows:

"MESSRS. JAMES LEFFEL & CO.:—I desire to build a mill-dam across a hollow about 60 feet wide. Will have a slate rock foundation all the way across, and the hight of dam will be 15 feet. I have two plans for building the dam. The first is to dovetail posts in the rock about four feet apart, straight across the hollow, and nail two-inch planks to the posts, setting the posts 12 inches in the rock, and having them 15 feet high.

My second plan (I think the best) is the following: Get me a sill 14 by 14 and lay on the rock across the hollow, and then put iron stirrups across the sill about 5 feet apart, placing the ends of the stirrups about 1½ feet deep in the rock and running Babbitt metal in the holes around the stirrups. Then place my posts about four feet apart, letting them about one foot deep into the sills, and having a brace running from top of post down stream, lower end on a sill. Stone is too scarce to build a stone dam. The stream of water is only from a large spring (no creek) only 100 yards from the dam. The water, when running on mill, is about 6 inches deep in a fore-bay 3 feet wide, affording enough water to run a set of 30 inch wool cards and a grist-mill, rocks 30 inches in diameter.

The fact is this: we want to build a good dam, without a great expense. I would say that the hollow is wider between

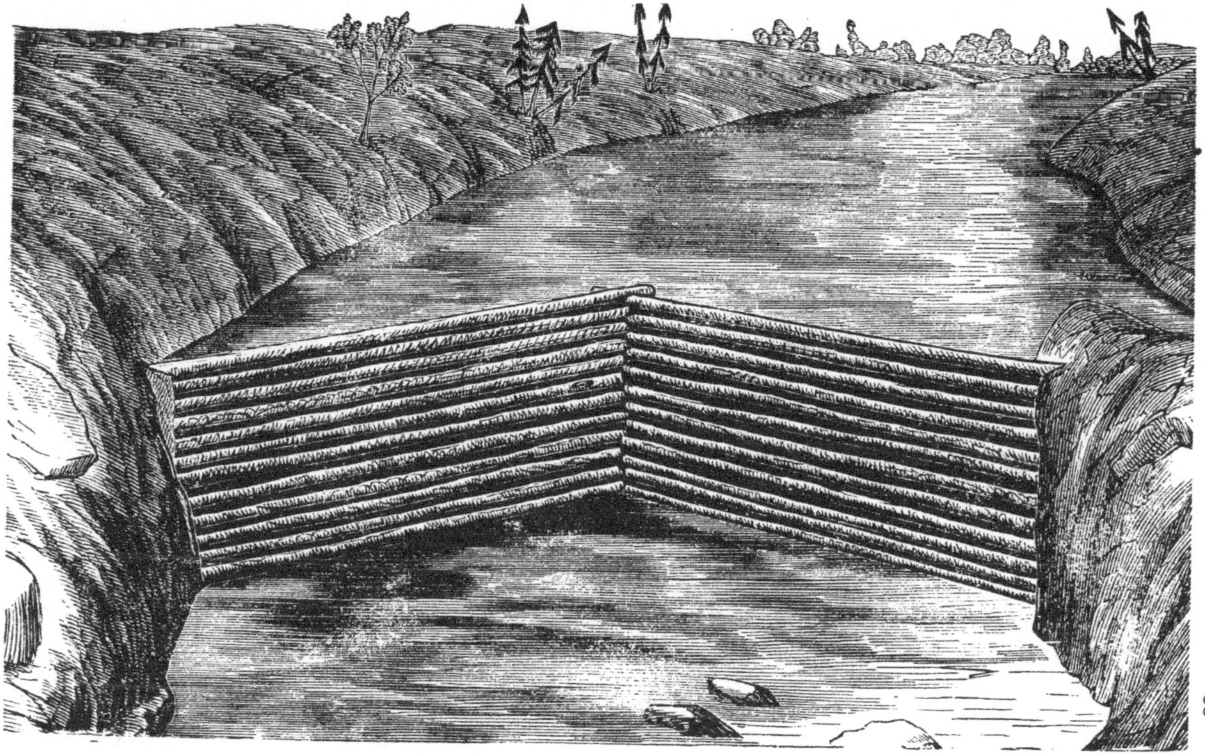

LOG DAM FOR NARROW STREAMS.

the spring and dam than it is at the point we wish to put the dam.

I desire you to answer this in the next number of the NEWS. Tell us which is the best plan, and if you can let us have a better plan, please give it in your paper. You may put me down as a permanent subscriber. Enclosed find money. Trusting you will comply with my request, I am

Fraternally yours W. H. W.

FLYNTVILLE, TENN."

"[We do not agree with our correspondent in thinking his second plan the best, but should give the preference to his first, provided certain important amendments are made to it. The posts should be let into the rock a depth of at least two and one-half feet, and three would be still better; and they should have one or if possible two series of braces if the dam is to be 15 feet high. The general plan of the dam, as we would build it, is similar to that described in our issue for November, 1871, except that the posts in this case should be nearer together than in the dam illustrated. The reason for this is, first, that the hight of the dam is much greater, and second, that slate rock is peculiarly liable to wear away under the influence of either sun, air, frost or falling water. We shall be glad to hear from other correspondents in regard to W. H. W.'s inquiry. ED. LEFFEL NEWS.]"

In a subsequent issue of the Mechanical News appeared a communication from another correspondent, in Wabash, Ind., over the initials "R. S.," giving a full description of a dam which he has found by practical experience specially adapted to narrow streams such as are here referred to. He also furnished a pencil sketch of a dam of this character of which he is a joint proprietor, located on Treaty Creek, Wabash Co., Ind.; and from the sketch thus supplied we have produced the engraving here presented.

Referring to the plans submitted by "W. H. W.," the Wabash correspondent remarks: "I do not like either of them; and as the editor stated that he 'would be glad to hear from other correspondents in regard to these inquiries,' I will, as an old hand at building mills and dams, suggest my plan of building dams across narrow streams, or 'hollows 60 feet wide' as he says his is. The motto of B. Franklin has ever been mine, viz.: 'What is worth doing at all is worth doing well.' Even should my plan cost a little more at first than his, it will be the cheapest in the end.

"Here it is. If the banks are stone, and have no natural jut or projection sufficient to abut the ends of the dam against, and are not too hard to cut, cut a groove in the stone embankments where the ends of the dam are wanted, about 12 inches wide, and a few inches deep, from the bottom up, and as high as the dam is to be built. Take round logs and face two sides straight and nice, large enough to measure a foot thick when faced. Cut the logs long enough for two lengths to make the dam. Square the end that is to go in the groove at the abutment, or shape it to fit. Lay the log not at right angles across the ravine, but put the ends which meet in the middle nine or ten feet up stream above a straight line, so as to form the dam that much in the shape of a horse shoe, or rather in the form of two panels of rail fence with the lock up stream; then halve together the ends which meet, putting the faces of the logs together as the dam is raised so as to hold the filling of gravel or dirt. Continue to so notch the logs together in the middle until the dam is the desired hight, filling up at the same time with gravel and stone if plenty; if not, dirt will do, provided the logs fit well enough to hold it. Thus we see, to build a dam in this way supersedes the necessity of any posts or braces, for it braces itself. And the harder the pressure of water and filling above, the tighter it will press the ends of the dam against the abutments, so that it can neither push out, wash round the ends, nor wash or undermine if stone bottom, and the bottom log well fitted. This plan supersedes, also, the necessity of cutting any post-holes or mortices in the bottom of the stream, or of bolting down the bottom log to keep the dam from pushing down stream."

"If not stone bluffs, then of course, abutments of either good stone or timber must be made, projecting into the banks. They should be notched up as the dam is raised, and all well filled as it goes up."

"If the bottom is slate, or any material that will not stand the force of water pouring over for many years, it should be leveled a sufficient distance up and down the stream, clear across, to receive a log apron. Face the logs on three sides, putting the square edge down. Cut them 16 feet long, and cut a gain on the top of each one 6 feet from the end that lies up stream, 4 or 5 inches deep, to lay the bottom log of the dam in, thus letting the apron extend about 10 feet below the dam, and 5 above. The apron logs notched in this way and the dam built on them, and they fitted up together, will prevent the bottom from wearing as long as they last. And having them 12 inches thick (which they should be for 15 feet fall) they will last, if water is kept over them, many years, for they cannot wash out put in in this way, nor raise at the lower end in case of a flood of water rising over them below the dam."

"This plan of building dams is not only applicable to W. H. W.'s 'hollow,' but to all streams that are not too wide for two logs to span in a bracing way. And it makes no difference how high the dam is built, it cannot push out if the logs are stout enough not to bend edgewise and come out like a spring-pole."

"We, of the firm of Small & Son, have a log dam 15 feet high, about 60 feet whole length, built precisely as I have directed, and it has been in use some twenty-seven or twenty-eight years, and not a log amiss yet; though the top is getting a little tender, and wants a new top-log. Ours are stone bluffs and solid limestone bottom, all the apron it needs. Our filling is mainly shelly limestone with some gravel and dirt; not even sheeted on top, but would be the better of it, for the stone and gravel washes off some."

"I give a rough sketch of our dam, which is so simple that any ten-year-old boy of common mother wit can see into the philosophy of its strength and durability."

"In 1846 I helped put a log dam across the Missisinewa River, nearly 200 feet long. In 1854, I think, I put in, or rather spliced the same kind of a dam on Deer Creek, to run a saw and grist mill. And about 1858 or '59, I put dam in to run two 4 feet burrs; all in Grant county, Not one of these dams has gone out yet, unless they this winter. Though these dams were all straight, breast is logs, and a log laid in the stream a few feet ab with dove-tail ties in it and the breast logs, as they notched up, all tied to the single log above, filled stone and gravel, then sheeted with 2 inch plank, graveled on the upper end of the sheeting; and with abutments and aprons, I consider the log dam the chea yet."

CHAPTER XXXIII.

FRAME DAM ON MAD RIVER.

We add in this chapter another illustration and description of a dam which has the advantage, over a merely theoretical plan, of being verified by actual construction, so that every detail has been worked out and may be relied on as practical, and duly adapted to the circumstances of the case. The dam here represented is built across Mad River, in Clarke County, Ohio, and is 165 feet long. The stream at this point has a mud and clay bottom, upon which is a coating of sand and gravel, washed down from above. The foundation of the dam consists of sills 30 feet long, hewed flat on the top and bottom to a thickness of ten inches, and laid lengthwise of the stream, about 8 feet apart. Upon the top of these sills, at their up-stream ends, and running across the stream, is bolted a timber 8 inches square, 16-inch bolts being used to secure it to the sills. The breast of the dam is raised to a hight of 30 inches above the apron, and is made by first laying a timber, 5 by 14 inches in size, across the stream on the foundation sills. On the top of this 5 by 14 inch sill are eight tiers of joists 3 by 10 inches, which are laid flatwise upon each other and spiked together with 6-inch iron nails. The face of the dam, composed of these joists, is battered or inclined up stream 5 inches. At a point five feet up-stream from the cross-sill on which the joists rest is laid across the stream, and bolted to the foundation sills, a timber 8 inches square. From this sill to the top of the breast-work of joists, or crest of the dam, are laid rafters 6 by 8 inches, 6 feet long, and 3 feet apart from centre to centre. The cut required at the upper extremity of each rafter to give it a secure hold upon the breast of the dam, is made about 3 inches from the end of the rafter, which therefore projects that distance in front of the tier of joists, and by this means the rafters are, so to speak, hooked over the crest of the dam. The depth of the cut is about half the thickness of the rafter, and the width 10 inches, the same as that of the joists. The rafters are pinned to the crest of the dam, and also to the sill at their lower ends, with wood pins 1½ inches in diameter.

For the covering of the dam there are laid, crosswise of the stream upon the rafters, 2½ inch planks, which are fastened to the rafters with 6-inch nails. From the foot of the rafters, also, to the up-stream end of the foundation-sills, a covering is laid consisting of planks 2 inches thick, running cross-wise of the stream and nailed to the foundation-sills. Upon the plank covering of the mud-sills, and extending some distance up the covering of the rafters, is a filling of gravel about 2 feet in depth; and the space under the rafters, from the mud-sills up to the plank covering, is also filled with sand and gravel.

The apron of the dam is made by laying three sills across the stream, resting on the foundation sills, and secured to them with 16-inch bolts. Upon these cross sills and the projecting edge of the 5 by 14 inch sill under the breast of the dam, are spiked planks 2½ inches thick, 12 feet long, and running lengthwise of the stream, as indicated in the engravings. At the down-stream end of these planks, against the side of the apron-sill and the ends of the foundation or mud sills, are driven spiles 3 by 6 inches, reaching to a depth of five feet. The same is done at the up stream end of the dam, where the extremities of the mud-sills and the side of the cross-sill at that point rest in like manner against the spiles. The spiles at both the up-stream and down-stream extremities of the dam, are placed close together, forming a continuous sheet across the stream.

The abutments of the dam are of solid masonry, laid up with cement, and are 21 feet in length of face, 5½ feet in hight, and 6 feet thick. In addition to this are the wings, each 10 feet long, and of the same hight as the face wall The tops of the abutments are at about the same level as the earth banks of the stream. They rest on the foundation

24

FRAME DAM ON MAD RIVER.

sills of the dam, three of which are under the abutment. The filling of the space enclosed by the face and wing walls is entirely of gravel and sand.

E. B. Harvey, of Miami county, is the builder of this dam, which is located near Enon, Clarke County, Ohio, and is owned by Martin Snyder. It supplies the power to run a flouring mill, propelled by Leffel Double Turbine water-wheels.

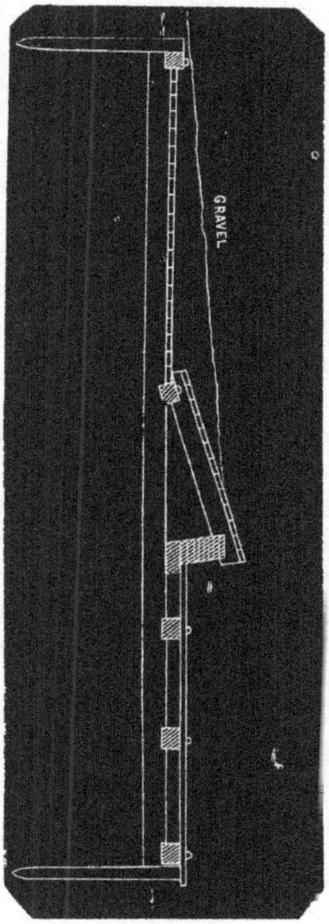

FIG. 2.

Our large engraving gives a perspective view of the dam, showing both the abutments and also a portion of the race, with the head-gates, three in number. In the smaller illustration is presented a sectional view, showing a foundation sill lengthwise of the stream; the end of the cross-sill at the up-stream extremity of the dam, and also of the cross-sill at the foot of the rafters, the one on which the breast-work rests, and the three sills of the apron; the ends of the eight tiers or layers of joists, and of the planks covering the rafters and the up-stream portion of the mud-sills; the side of one rafter, showing the cut at one end for the foot-sill and at the other end for the crest of the dam; the plank covering of the apron, the spiles at the upper and lower extremities of the dam, and the line of the gravel filling. By the clearness with which every detail is shown in this cut the whole construction of the dam will be accurately understood.

We may here state that for the purpose of keeping the water from interfering with the work upon the dam, a coffer was built at a point 5 or 6 feet up stream from the upper ends of the foundation sills, extending from one bank nearly across the stream, and thus protecting one-half of the dam while the building was going on. To protect in like manner the other half of the dam while in process of construction, it was only necessary to remove the upper part of the coffer above the finished part of the dam, letting the water flow over both coffer and dam on that side of the stream; the material thus taken off being used to extend the remaining part of the coffer to the opposite bank, and a barrier being also built from the coffer to the inner or midstream end of the finished half of the dam, keeping the water from that part of the dam on which work was still in progress.

[Since the foregoing was put in type, we have received information that the work upon this dam, just as it was approaching completion, was interrupted in the following manner. The dam on one side of the stream having been finished, and work on the remainder being in progress under the protection of a coffer, a channel was cut from the head race through the bank behind and around the finished abutment, to carry off the water. A flood occurring, the swollen stream poured through this channel and caused great damage to the abutment and the completed portion of the dam. We presume that parties on the spot were best qualified to judge as to the course proper to be pursued; but from what data we possess, we are inclined to believe that by first putting in a head-gate at the race, and allowing the water to pass over the completed portion of the dam, the disaster might have been avoided.]

CHAPTER XXXIV.

DAM AT OSBORN CITY, KANSAS.

The dam herewith illustrated is constructed on the same principle, in many respects, as several which have already been described; but the plan here shown will be found in some localities to possess advantages in point of simplicity and strength which will justify its adoption by the mill-owner. It cannot be accurately classified as regards the kind of material employed, as stone, logs, sawed timbers, boards, rock, gravel, sand and hay are used in its construction, their proportions and arrangement being such as to afford, without very heavy outlay, a satisfactory degree of firmness and durability. Our engraving gives a perspective view of the dam built by Messrs. David Milne & Son, at Osborn City, Osborn Co., Kansas, furnishing power by which their saw-mill and grist-mill are run. The width of the river at this point is 64 feet, and its bottom consists of a layer of sand about 3 feet in depth, resting on a bed of solid slate and shale. In preparing for the erection of the dam, the first step was to scrape the sand away until the solid bottom was reached. The mud-sills were then put down, consisting of logs from 14 to 18 inches in diameter, and from 20 to 28 feet long, their direction being lengthwise of the stream. Brush was also put in to aid in making the sills as firm and solid in their positions as possible. In scraping away the sand, a hollow of considerable depth was of course made; and after the sills were put down, the sand washed over the mud-sills, which thus became imbedded in sand and brush, and have thus far shown no indications of giving way. The distance between the sills is about 6 feet from center to center.

It should be stated, before proceeding further, that the

body of the dam is supported at each end by wing walls, as shown in the cut, these walls being 3 feet in thickness and built solidly along the face of each bank for a considerable distance both above and below the dam.

After laying the mud-sills, as described, the next stage of the work is the erection of the crib, which is composed of sawed timbers, and rests upon the sills, extending from bank to bank, and forming, as will be seen in the engraving, an obtuse angle with the vertex up-stream. The width of this crib is five feet and its hight 8½ feet. The timbers running across the stream are 6 by 6 inches, while the cross pieces are 4 by 6 inches, placed flatwise, from 5 to 7 feet apart, and spiked to the main or longitudinal timbers, which are therefore 4 inches apart, one above the other. These 4-inch spaces are covered by nailing boards upon them, thus rendering the up-stream and down-stream walls of the crib sufficiently tight for all practical purposes. The 6 by 6 inch timbers are pinned together with 2-inch oak pins, 16 inches long. In the engraving, the dam is shown with a part of the filling on the down-stream face cut out, giving a view of a portion of the crib in the interior. The main or longitudinal timbers, the ends of the cross-timbers, and one of the foundation sills, are thus shown, and also the level top of the crib, 5 feet wide, forming the crest of the dam. The ends of the foundation sills are likewise seen, projecting down-stream from under the filling. Each end of the crib, at the point where it joins the wing wall, is let into the wall for a depth of 3 or 4 inches, giving it a firm and solid bearing, and rendering it, in connection with the angular direction of the two halves of the dam, abundantly strong in its position, so far as regards any direct pressure from the water above.

The filling of the inside of the crib consists of broken rock, gravel and hay, arranged in the following manner: a layer of rock, finely broken, is first put down, having a depth of 10 inches; a coat of gravel is then put on, leveling up the surface of the rock; then follows a layer of hay, then another layer of rock, and so on with alternate coats of rock, gravel and hay up to the top of the crib. The rock

DAM AT OSBORN CITY, KANSAS.

used is a kind of flint found in the vicinity, and very heavy. Above the crib is a filling of broken rock, gravel, hay and sand. The width of this fill at the base is 12 feet, sloping to the top of the crib. Below the crib, on the down-stream face of the dam, is a fill of rock and brush, sloping to the top of the crib, and the whole dam has thus the shape of the roof of a house. The crib is located at a point on the mud-sills about two-thirds of the distance from their down-stream to their up-stream extremities, and the front of the crib is just above the projecting corner or vertex of the angle formed by each of the wing walls. These distances and proportions are distinctly shown in the engraving.

This dam has been doing service for some time, and as yet has shown no signs of settling, or of deficiency in any respect.

We are of opinion that the plan of dam above described, which is an excellent one in most respects, would be still further improved by bolting the mud-sills in a few places to the rock bottom. If they were surrounded and covered by a good depth of mud, this would be less important. A mixture of sand, in liberal proportions, with the gravel in the crib, to pack and tighten the whole mass, would also be useful; although this point is very well provided for by the board covering on the side of the crib, especially if a considerable amount of fine sand and gravel is thrown against it. As for the use of hay, either in the crib or above it, we have small faith in its utility, as it will rot out after a time and require refilling. There is, in fact, nothing better than heavy gravel and sand for all kinds of filling about dams, head-gates, races, &c.—and nothing poorer than clay.

Of the light rocks and brush forming the inclined apron below the dam, a considerable portion will wash away in case of a flood; but if there are also plenty of heavy boulders, these will maintain their position, and no material damage will be done.

CHAPTER XXXV.

STONE AND TIMBER DAM.

In the present chapter we describe and illustrate a stone and timber dam which was erected in 1873 by the owner, Hamilton B. Lawton, at East Brunswick, Rensselaer Co., N. Y. Its method and material of construction are such as to adapt it to a region where stone is abundant, as this, with a moderate amount of timber, is the article principally used in its erection. This dam is built on a rock and "hard-pan" bottom. Its length is 150 feet, and its hight 22 feet, from the level of the water to the top of the upper plate. The base of the dam, measured on a horizontal line from the up-stream to the down-stream extremity, is 23 feet in extent, being nearly the same as the hight; and the up-stream side of the dam, therefore, slopes at an angle of 45 degrees. This form of construction gives the necessary degree of stability, and also affords ample room for filling in between the rafters with rocks and small stones, thereby rendering the mud-sill and plate very secure in their position.

Our principal engraving shows the face of the dam and abutments, the upper and lower plate and the posts being the only timbers visible. In the smaller cut is given a complete representation of the framework, in which A is the upper plate and E the mud-sill at the up stream extremity of the dam. The rafters B are fastened to the plate and sill with strong spikes. It will be observed that midway of the rafter B is a timber D, parallel with the plate and mud-sill; and that to this middle plate are attached short rafters C, alternating with the main rafters and having their lower ends secured to the up-stream mud-sill in like manner with the main rafters. The purpose of these short rafters is to

STONE AND TIMBER DAM.

give a more firm support to the plank covering of the dam at this point, where the pressure of the water is heaviest.

The main rafters, reaching from the up-stream mud-sill (which is bolted to large rocks) to the upper or cap plate of the dam, may consist of timbers unhewn except on their upper faces, where they should be made flat to admit of the laying of the planks, and give an even surface to the water side.

The main timbers of which this frame is composed are 12 by 14 inches. The bottom plate or sill F lies upon a series of rocks arranged, as indicated in our first illustration, so as to form an apron to receive the overfall of water and pre-

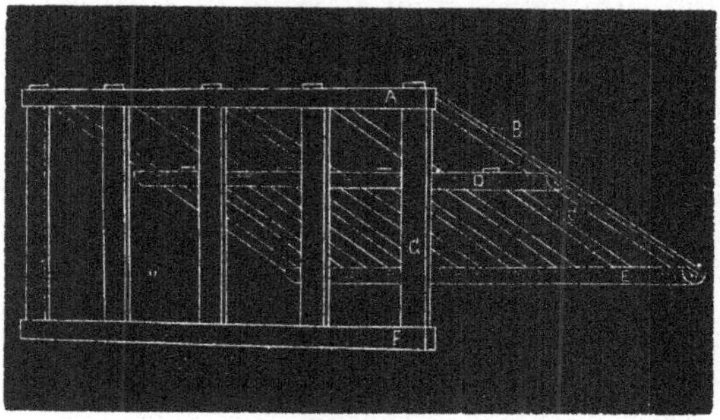

FIG. 2.

vent the washing, wearing and undermining of the base of the dam. The posts G are framed into the upper and lower plates as shown in both the cuts, and serve to support the upper plate, in case the wall should settle in course of time, as it is liable to do to a very small extent.

The ends of the dam, on each side of the frame-work, are compactly built up with rocks and small stones in the rear, and in front square rocks are carefully laid up to present a smooth front and a permanent wall; thus allowing the timber work to be taken out and renewed, should it be necessary at any future period.

The filling in the interior of the framework, as already mentioned, is composed of rocks of irregular size, from heavy boulders down to cobble-stones; and the dam is covered with planking in the same manner, substantially, as described in former chapters relating to dams of this general nature.

The builder of this dam is confident that it will last a lifetime, and that very little expense will be required in repairing the wood-work. The other portions of the structure should of course demand no outlay whatever after having been once completed.

As an appropriate addition to the account above given, we may here describe briefly another dam of very similar nature to the one already shown, although in quite another section of the country. The dam to which we now refer is built across the Des Plaines river at Joliet, Ill. It is the lowest dam down the river in the city, there being two State dams above it; and is owned by Messrs. Wm. Adam & Co. of the "City Mills." The bottom of the river at this point is limestone. The dam has an extent across the stream of 160 feet. Its face is composed of masonry, with the addition of a mud-sill and cap-sill, the whole corresponding almost precisely with the face of the dam already described, except that there are no upright posts connecting the upper and lower sills.

The mud-sills, which cross the stream, are 12 by 14 inches being of the same size as those in the other dam. The first mud-sill, at the face of the dam, is laid on the rock, which is leveled off as smoothly as possible to receive it; and 40 feet up-stream from this is placed another 12 by 14 inch sill, parallel with it, the two being bound together with timbers 6 by 8 inches, running lengthwise of the stream, and placed at intervals of 12 feet for the whole width of the stream. Around the first mud-sill a stone wall is laid in water-lime, and on the foundation composed of this wall and sill the face of the dam is built, consisting of solid masonry, 30 inches in thickness. On the top of this wall is placed the 12 by 14 inch cap-sill. This cap-sill or plate is kept in its place by means of binders 6 by 8 inches, which extend from the plate to the up-stream mud-sill. These binders are fastened to the timbers which tie the two mud-sills together (as already described) by iron rods, and are also supported by posts to give them the necessary stability. Furthermore, across these binders, which run lengthwise of the stream, smaller timbers, 4 by 6 inches, are framed, parallel with the face of the dam, to keep the binders from spreading apart.

All the timber work in this dam is dovetailed where cross-timbers are met; and in fastening the framework together, ten kegs of 8-inch spikes were used, from which it will be seen that it is not likely to become separated by any strain it is liable to undergo.

Back of the face of the dam a layer of clay was filled in, clear up to the face of the cap-sill; back of this, brush, rubble stone and gravel were put in; and on top of this was spread a coat of clay. A covering of two inch planks was then put on, the whole length of the dam, for a distance of 20 feet from the crest toward the up-stream extremity, comprising, in other words, the upper half of the up-stream slope, next to the face of the dam. Finally, a covering of gravel was spread over the entire up-stream slope, with the exception of about six feet along the cap-sill.

The banks of the stream are faced with a wall of masonry, connecting with each end of the dam and forming the abutments. There is certainly no lack of strength and solidity in the dam, and its manner of construction and selection of material appear to be, for the region in which it is located, of a very judicious character. If we were to take any exception whatever, it would relate to the use of clay as one of the materials for filling, our own experience and observation having convinced us that it is less reliable for this purpose than any other substance used, whether it be gravel, loam, sand or brush. Undoubtedly, in the case here described, the other sources of strength and compactness in the structure of the dam will preclude any danger of injurious results from the presence of the clay; and when thus protected by better material, it may answer as well as any to a limited extent; but in cases where it is expected to resist of itself the inroads of the water, we should not regard its use as safe or profitable.

CHAPTER XXXVI.

DAM FOR QUICKSAND BOTTOM.

In the issue of Leffel's Illustrated Mechanical News for November, 1873, appeared the following inquiry from a Kansas correspondent, over the initials "D. P.": "How can a dam be put in with 10 feet of sand or quicksand at the bottom?" In this inquiry is presented one of the most formidable difficulties with which the builder of a dam has to contend; and he is therefore fortunate if the nature of the river-bed on which he is to place his foundation is not of the character described in the question above stated. The trouble is not so much in devising a theoretical plan to meet the conditions of the case as in the lack of due consideration, in many cases, on the part of the builder, resulting in insufficient provision for the peculiar dangers to which the structure will be exposed. There is no case, in the prosecution of enterprises of this kind, where deliberate forethought, and careful study of all the circumstances and necessities of the situation, are more essential to ultimate success, than in the one which we have now in view. For this reason, it is especially desirable that the teachings of experience, rather than of mere theory, however plausible, should be consulted for the guidance of the builder; and we will not pause, therefore, to discuss on abstract grounds, the principles which should be followed in a case of this nature, but proceed at once to describe a dam in the erection of which the obstacles referred to were encountered, and which has shown by its permanence that it possesses all the necessary elements of durability.

Our engraving gives a general view of the dam to which we allude, viz.: the "Hydraulic Dam" across the Tippecanoe River at Monticello, White Co., Ind., which was built

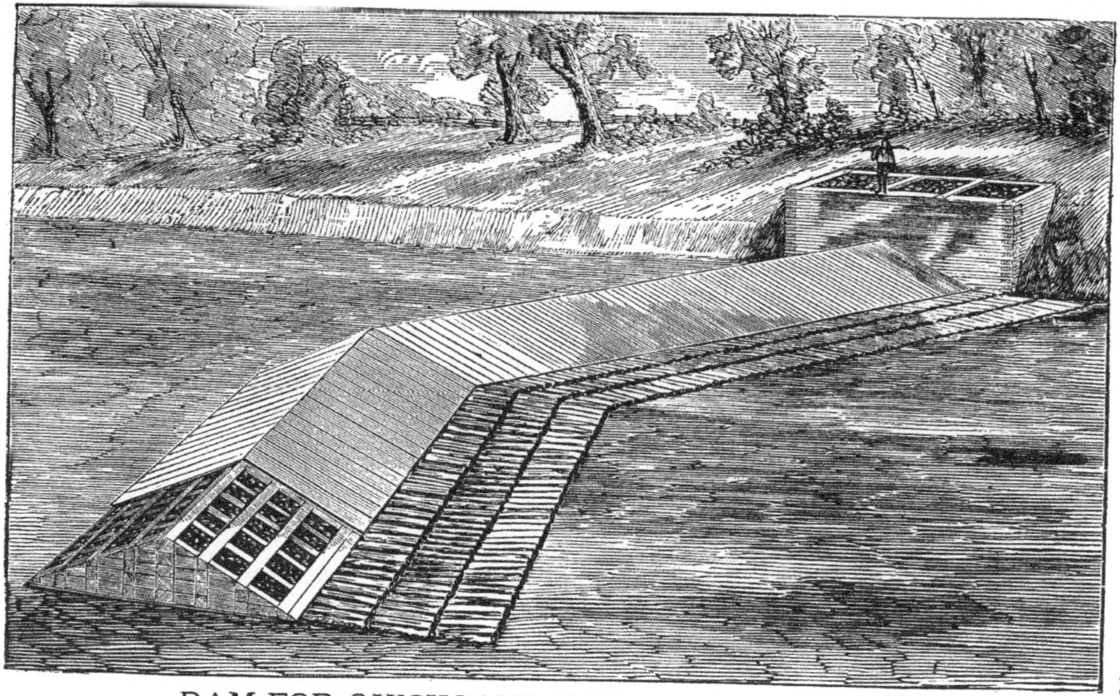

DAM FOR QUICKSAND BOTTOM.--Monticello, Ind.

m the year 1849, under the direction of Mr. E. A. Magee, for the Hydraulic Co., of that place.

This dam rests upon a quicksand foundation, and the banks of the stream on each side are also of sand. The length of the dam between the abutments is 340 feet, its width from up-stream to down-stream extremity (exclusive of the apron in front) is 24 feet, and its hight 5½ feet. The abutments, only one of which is shown in the engraving, are each 30 feet long, 12 feet high and 12 feet wide, and are composed of timber and rock as hereafter described. The foundation of the dam, part of which constitutes the apron, is laid as follows: commencing with the down-stream tier of the apron, the lower extremity of which is 18 feet from the down-stream edge of the main portion of the dam, poles or small trees from six to eight inches in diameter at the butt and from 40 to 50 feet in length, with all the brush left on at their upper ends, are laid lengthwise of the stream as close together as possible, and rock enough placed on them to hold them to their position. A second tier of the same kind is then laid, the ends of the trees being six feet back of those in the first tier; and a third tier follows in like manner, falling back from the second tier six feet as before.

Six feet up-stream from the ends or butts of this last tier of trees is commenced the base of the dam itself, which is thus already provided with a secure foundation, composed of the upper portions of the three tiers. As the entire distance from the up-stream extremity of the 24 ft. dam to the down-stream edge of the 18 ft. apron is but 42 feet, and as the trees in the three foundation tiers are 40 to 50 feet long, their upper portions will of course extend under the whole base of the dam. The weight of the dam serves to hold them securely in place, and they in turn give the dam a hold upon the bed of the stream of such breadth and strength that it is practically immovable.

For the first course at the base of the main dam, seven sills or stringers are laid, cross-wise of the stream, the one farthest down stream being, as stated, six feet from the ends of the uppermost tier of trees in the apron. The ends of these sills are seen in our engraving. They are 14 by 16 inches in size, and their distance apart, between centers, is 4 feet, dividing the width of the whole base into 6 equal parts. They are lapped on each other where two ends meet, and fastened together with two-inch pins. Their upper sides are counter-hewed to receive the cross-timbers, which are put in at intervals of 10 feet for the whole length of the dam. These cross-timbers, whose direction is of course lengthwise of the stream, are 12 inches square, and those in the first course are 24 feet long, being equal to the width of the base of the dam. They are counter-hewed and let down evenly on top of the first course of stringers or sills.

The second course of stringers, which are five in number, are 12 inches square, and counter-hewed on top like the first or bottom course, upon which they rest solidly, making, with the cross timbers, a "water-joint." Next comes the second course of cross-timbers, 12 inches square and counter-hewed, but shorter than the preceding course, the width of the dam being less as it approaches the top; then the third, fourth and fifth courses of stringers, alternating with the third, fourth and fifth courses of cross-timbers, all of which are 12 inches square, and counter-hewed so as to form the "water joint" by their contact wherever they cross. The fifth or last of the courses of stringers consists of a single timber 12 inches square, laid solidly on the center tier as shown in the engraving. The outsides of the outer stringers, and the ends of the cross-timbers, are beveled so as to present a smooth and even inclined face, which is planked on both the up-stream and down-stream slopes of the dam, as shown in the cut. The planks used are 2½ inches thick, and are fastened to the timbers with 6-inch spikes.

By the crossing of the sills and transverse timbers in the frame, with water-joints as above described, a large number of cribs are formed; and these are filled with rock up to the comb of the dam.

The abutments, the dimensions of which have already been stated, are composed of timbers 12 inches square, counter-hewed and laid solidly one upon the other. They are lapped at the ends and pinned with two-inch pins.

Through the interior of the crib thus formed extend two courses of ties as shown in the cut, dividing it into three smaller cribs, all of which are filled with rock to the top. Outside of the abutment, both up and down the stream, for a distance of three feet, the bank is excavated, and the sand thus taken out is replaced with fine gravel and clay and sand puddle. On the side and end next to the water above the dam, sheet piling is driven, and the abutment is planked up and down with 2-inch planks. On the upper side of the dam and on the brush of the tree-tops projecting above, a coating of gravel two or three feet thick is placed.

It will be observed that this dam by its construction forms an angle across the river, with the point or vertex up-stream, thus giving it to some extent the elements of strength pertaining to the arch, but requiring less care in the framing than if a regular curve was made across the stream. It should be here stated that the engraving is in some respects an imperfect one, as it does not show the planking on the abutment; and in the cross section of the dam in the front of the picture, the shape and arrangement of the counter-hewed sills and cross-timbers are not accurately represented. The cut is, however, sufficiently correct to enable the reader, aided by the minute description we have given, to form a clear idea of the manner in which the work is done.

In building this dam, 15,000 feet of hewed timber, 25,000 feet of plank, and 1,575 poles or trees were used. The total cost at that time (1849) was about $4,500, but would be greater now, as labor and materials are both more costly than 25 years ago. The durability of the structure, with its broad base and the pyramidal form of the main dam, are sufficiently manifest, the strength of the abutments and the weight of the filling both in abutments and in the cribs of the dam, being such as to give abundant stability, in spite of the unfavorable nature of the river-bed. The fact that the dam has stood for over 24 years, and that the main body of it is still perfectly good, is conclusive testimony on this point. For a period of 15 years it required no repairs; but afterward the abutments above the water were rebuilt, and some repairs have recently been made on the dam itself.

CHAPTER XXXVII.

OVERHUNG APRON DAMS.

Dams of several different kinds, adapted to streams having a hard bottom, have been illustrated and described in preceding chapters of this work; and as the one of which we now propose to speak does not differ greatly from some already shown as regards the material used and the general principle of construction, we have given only an outline sectional view of it, which will, however, present with sufficient clearness all the novel features contained in it. The apron, in fact, is the only point in which there is any radical departure from the system laid down in former chapters. In this respect, the dam here shown is quite peculiar; but as it has stood the test of practical service for a number of years, we must conclude that for the locality and the kind of foundation on which it is built, it is a reliable structure, at least under any but the most extraordinary circumstances.

This dam was built in 1867, Mr. C. Goodnow of East Sullivan, N. H., with other parties, performing or superintending the work. Its hight is 13 feet, and its length about 60 feet across the bed of the stream, at the point where the foundation timbers lie; while, measuring on the cap or top of the dam, the distance is 80 feet. One end of the dam rests against a ledge, while at the other end is a steep gravel bank.

In the cut, (Fig. 1) A A represent the foundation-sills extending across the stream, which consist of whole trunks of trees, some 24 inches in diameter at the butt. These sills are made flat on their upper surface to receive the cross-sills B, which are 12 inches square and locked on the top of the sills A, as shown in the cut, the gains being two inches in depth. The spaces between the sills are filled with rocks. The rafters C are 28 feet in length, 12 inches in diameter at the lower end, and 10 inches at the upper end. They are notched upon the up-stream sill A, and at the other end upon the cap-timber F. The posts D are 10 inches in diameter, with a 3-inch tenon at each end, one being inserted in the cap-timber F, and the other in the cross-sill B, which runs lengthwise of the stream and resting on

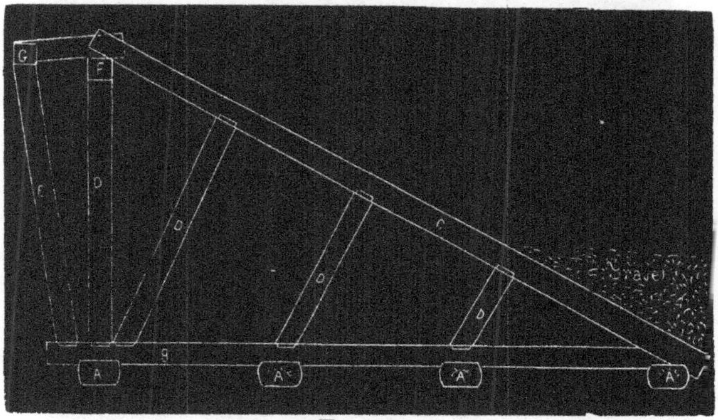

Fig. 1.

the foundation-sills A. The length of the posts D is sufficient to give the dam, at the crest, a hight of 13 feet, as already stated. The distance between the rafters is 4 feet, from center to center.

The manner in which the apron is framed is plainly indicated, and will attract the particular attention of the reader. The sills B project down-stream beyond the front foundation-sill A far enough to receive the posts E, which incline somewhat down-stream from the front of the dam. At the upper end of the posts E they are framed into the cap-timber G, from which short timbers extend to the top of the cap F, meeting there the upper ends of the rafters. The projection thus formed, which we have called the apron, (although it bears but little resemblance to that portion of the dams hitherto described) serves to carry off the water from the dam, the overflow in floods being frequently 20 to 30 inches deep on the crest. Of course, on any other than a rock or very hard gravel bottom, an apron of the usual kind, and of considerable extent down-stream from the base of the dam, would be required to prevent washing and undermining; but there appears to have been no necessity for it in this case, as the builder of the dam states that it has

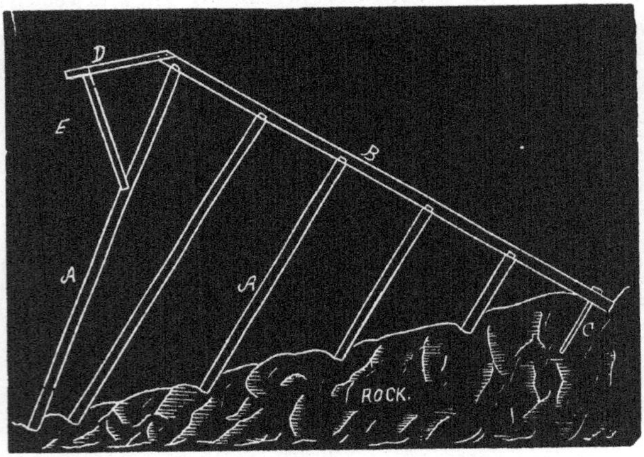

FIG. 2.

withstood all the floods which have occurred in the period—reaching nearly seven years—which has elapsed since its erection.

It should be further stated that the rafters of this dam are covered with planks 2½ inches thick, which are secured to the rafters with 5-inch spikes. The total amount of lumber used in building the dam was 25,000 feet.

Another dam, strongly resembling this in its method of construction and in the kind of apron attached, but of a still simpler form in many respects, is in use at Millbrook, Dutchess Co., N. Y. We give in Fig. 2 a representation of the manner in which it is built, from which it will be seen that nothing more simple or economical in the way of frame-work can well be devised. The particulars in regard to this dam are furnished to us by Mr. V. Anson of Millbrook, who states that the stream on which it is built has a rock bottom and sides, the river bed being quite steeply inclined, as indicated in the cut. No mud-sills are laid, and no timbers are required to rest the braces or studs A A upon, as they are footed directly into steps or notches in the rock. The distance between these studs is 5 or 6 feet, or whatever space may be adapted to the hight it is desired to give the dam. In framing the studs into the rafter B, the builder of this dam states that he found it much better not to make long tenons, secured with pins, as the timber would give out in and around the tenons. He therefore made them quite short, just enough in fact to keep them firmly in their places, and omitted the pins; and the results were entirely satisfactory.

The studs A and the rafter B constitute (aside from the apron) one bent of the dam; and these bents are placed side by side in a direct line across the stream, with intervals of 2 or 3 feet between them. Having been set perfectly plumb and properly stayed, they are covered with planks two-inch pine being considered sufficient for this purpose

At the foot of each rafter, up-stream, an iron rod C, 1¼ inches in diameter and 2 feet long, passes through the rafter and into the rock for a considerable distance. It is manifest from the position of the rafter and direction of the rod that the pressure of the water from above the dam will tend to keep the rod in its place rather than to withdraw or loosen it; and it will have a like effect to preserve the foothold of the studs in the notches cut for them in the rock bottom.

The cut shows very clearly the construction of the apron, the timber D having a slight incline from the horizontal, and being halved on to the end of the rafter and secured by pins or otherwise; while near its outer extremity it is supported by the brace E. One end of this brace is tenoned into the timber D and the other into the stud A, as shown. It will be observed that the stud A at the front of the dam is not parallel with the others, but is drawn in at the foot—the object being to avoid the fall of water from the apron upon the foot of the studs, which would in time loosen them and weaken the entire structure.

The studs and rafters are 12 inches square, and the timbers of which the apron is made 4 or 5 inches square. It is unnecessary to remark that while this form of dam may be entirely reliable on a rock bottom, with banks of the same character, and other circumstances of a favorable description, it would be impossible to give it the requisite strength and firmness, on the majority of streams, without mud-sills, and also an apron at the front of the foundation, such as we have illustrated in former chapters.

CHAPTER XXXVIII.

STONE DAM WITH PLANK COVERING.

It will be perceived on a very superficial examination of the dam illustrated in the present chapter that it is of an extremely substantial nature, and presents no weak point in any part of its structure to lead to a destructive inroad of the current. The plan of the particular dam shown in our engraving is taken from drawings furnished us by Messrs. Fassett & Stevens, of Lewiston, Maine; this dam having been built on the Sabbatus River at Lisbon, Maine, for Hon. N. W. Farwell. The same parties have built a number of dams of the same general description, it being adapted to any locality where the river has a ledge bottom and sides, and in such cases has never failed to give entire satisfaction. For any other sort of bottom or banks it would of course be unsuitable, without very material modifications, and the addition of abutments, apron, &c., none of which are here required.

The dam here referred to is about 150 feet long and 10 feet high. The body of the structure, as will be seen, is simply a solid wall of masonry, of the hight above stated, about 14 feet broad at the base, perpendicular on the down-stream face, and sloping on the up-stream side to a breadth of some six feet at the top. Resting on solid rock, it requires no foundation sills; but a bed-sill or timber is provided to receive the lower ends of the planking, and other timbers are placed at suitable points, as shown, for attachment of the planking in a perfectly secure manner. All these timbers run across the stream, lengthwise of the dam for its whole extent. The bed-sill first mentioned is

12 by 12 inches in size, and is placed, not directly upon the ledge, but at an elevation of about two feet, resting on an intervening foundation of brick built in pyramidal form as regards its cross-section, and laid in hydraulic cement. The purpose of this is to give the bed-sill a perfectly water-tight footing, as there might otherwise be a possibility of leakage beneath it which would loosen the timber in its position and consequently impair the security of the planking. The bed-stick is secured to the ledge by means of pins of round iron, 1¼ inches in diameter, and of sufficient length to enter the ledge, below the brick-work, at least 15 inches. These pins are placed at intervals of not over five feet between their centers, requiring in all 30 pins for the whole length of the timber. They are well driven, and wedged at the lower end to prevent any liability to work loose. The manner in which this is done is indicated in the engraving, and has also been mentioned in a previous chapter; the end of the pin being split five or six inches up and an iron wedge inserted, which, when the pin is driven down, comes in contact with the rock and is thus forced up, spreading the point of the pin and giving it such a hold as to prevent the possibility of its withdrawal. The introduction afterward of fine wet sand will give the desired firmness to the pin, and is recommended for this purpose as equal to lead or cement.

The other timbers are of the same size as the bed-sill, 12 by 12 inches, and run in the same direction. The manner in which they are secured in the stone-work is plainly shown in the illustration, pins of the same material and size being used as on the bed-stick, but with the ends turned at a right angle, it being impracticable to wedge them as in the other case. The timber at the front of the dam is square, while the other two, as will be seen, are beveled in such a way as to adapt them to the planking. The planks of which the covering consists are 3½ inches in thickness, and laid as indicated in the cut.

The stone for such a dam as is here described may be taken from any ledge of field stone. About one foot of the thickness of the dam, on the face or down-stream side,

STONE DAM WITH PLANK COVERING.

should be laid in cement mortar, to ensure its durability and tightness.

The shores being each a solid ledge, no abutments are required, it being only necessary to make the connection of the dam with the ledge perfectly tight by means of the cement and filling. The filling, on the up-stream side, may be of dirt, sand and gravel, its only purpose being to exclude the water from the base of the dam.

The apron of this dam, it will be observed, is a natural one, the front of the wall being within two or three feet of the edge of a steep ledge about four feet in hight, extending across the stream.

Hon. N. W. Farwell, for whom, as above stated, this dam was built, is the proprietor of very extensive cotton mills, and is using several of the larger sizes of the Leffel Double Turbine water wheel.

CHAPTER XXXIX.

TIMBER DAM AT SOUTH HADLEY FALLS, MASS.

One of the largest dams ever built in the United States is that which extends across the Connecticut river at South Hadley Falls, 8 miles north of Springfield, Mass. It was completed in Oct., 1849, the work having been prosecuted by the Hadley Falls Company, incorporated for this purpose in 1848 with a capital of $4,000,000.

This dam is 1,017 feet long, and 28 to 32 feet in hight, and is built of timber, with the exception of the interior filling and the abutments, the latter being of solid masonry. Before laying the timbers, the river-bed, which is of solid rock, was excavated to a depth of 4 feet, and a width up and down-stream of 90 feet, being equal to the base of the dam. The first sill was then laid, 12 by 12 inches in size, extending across the stream, and bolted down to the bed-rock with 2-inch iron bolts. Rafters of the same size were then placed lengthwise of the stream, two feet apart, extending from the sill to the rock bottom, and sloping up-stream, their ends being scarfed to fit the bottom, and secured with 2-inch iron bolts at both ends Timbers were then laid cross-wise of the rafters, 2 feet apart, followed by another set of rafters, and so on until the desired hight was reached. The work was protected while in progress by a coffer-dam. The size of the timbers throughout was 12 by 12 inches, and all were fastened with 2-inch bolts as already described.

The spaces between the timbers were filled in with stone for 15 feet from the bottom, and gravel was laid over this and in front. The slope from the top of the dam to the upper edge of the base is 21½ degrees. The covering consisted of 6 inch planks, bolted to the timbers, the ridge being double planked; and at the points where it was most exposed to damage by ice it was further protected by a covering of boiler-plate iron. About 4,000,000 feet of timber was required for the construction of this dam, in addition to the large quantity of stone for the abutments, as well as for the filling at the foundation and between the timbers.

The piers, as already stated, are of solid masonry, but the strain upon them is in fact much less than might at first sight be supposed, as the dam, which is built straight across the stream, is so constructed as to be practically self-supporting. It has in fact withstood the heaviest freshets ever known in the Connecticut river. A striking proof of its reliability was afforded in the great October flood which occurred a few years since, when the stream pouring over the dam measured in depth, on the cap of the dam, twelve feet, three inches.

When the dam was built, the rock bottom was deemed sufficient to protect the structure from being undermined by the overfall. About two years ago, however, the rock under the dam was found to be undermined by the action of the water (when impeded by ice that had gone over the dam and piled up in the stream below) so that the whole fabric was in danger of sinking. About one million feet of lumber and stone was required to fill the excavation made by the water, and build an apron sloping from the top of the dam down-stream, its total extent being about three hundred feet.

The water is admitted by 13 gates to a main canal faced with masonry, 140 feet wide at bottom, 144 feet at the top, and 22 feet deep, branching at a distance of one-fifth of a mile from the river into two mill-races, for the use of factories on different levels. The water from the upper race, after moving the mills on its proper level, is conveyed back to a point near the river, where it falls into the lower race. The motive power afforded by this dam is believed to be without a superior in this country. It is utilized in the propulsion of extensive cotton-mills, paper-mills, &c., the products of which reach a formidable yearly aggregate, and represent the labor of several thousand men. The dimensions of the dam, and the volume of water which is thus made available for manufacturing purposes may be judged from the fact that the roar of the fall is sometimes heard for a distance of 40 miles; and at Springfield, which is eight miles distant, doors and windows are frequently observed to rattle in unison with the vibrations produced by the overfall at the dam, and of course more distinctly perceptible in its immediate vicinity. The occurrence of these vibrations has afforded an interesting subject of study to scientific men, it being found that they vary with the varying temperature of the atmosphere, increasing in number as the temperature rises. When the thermometer indicated 70 deg. Fahrenheit, the vibrations were 130 per minute; at 80 deg., they rose to 137 per minute. The observations of which this report is made were taken by Prof. Snell of Amherst College. The general subject of vibrations of this character has also been investigated and discussed by Prof. Loomis, who states that the vibrations have in some cases been so marked and continuous as to be a source of extreme annoyance to persons living in the vicinity; and it will readily be comprehended that the monotony of such an effect, when indefinitely prolonged, would become almost intolerable to people of very nervous and sensitive organizations. It appears from a number of instances on record that the vibrations do not occur, or at least do not exhibit the regularity otherwise attending them, unless the water falls in an unbroken sheet; and they have been interrupted or altogether stopped by a floating log catching on the top of the dam, or by strips of wood attached to the crest of the dam expressly for the purpose. The question whether the air or the earth is the medium of transmission of the vibrations is an unsettled one, although the powerful influence of the temperature of the air upon their frequency would seem to favor the theory that they are principally conveyed by that agency. It is quite possible, however, that both the air and the ground, (especially if there are extensive intervening ledges of

, affording a continuous transmitting medium,) may e in this instrumentality.

e are indebted for many interesting facts relative to dam above described, to Mr. M. H. Arnold of West ckbridge, Mass.

CHAPTER XL.

STONE APRON DAM.

In the dam which we illustrate in this chapter, the usual order of construction appears to have been reversed, and what would at first glance be taken for the up-stream is in reality the down-stream slope; while the perpendicular side of the dam, instead of being at the front, with an apron at its foot, is at the back of the dam, the whole remaining part of the structure serving in the capacity of an apron, and carrying off the water in a gradual fall or rapid. But notwithstanding this reversal of the method ordinarily adopted, the dam here shown is believed to possess great durability, and for the locality in which it was built is said to be extremely cheap. An examination of the cut is sufficient to convince the practical reader that provided the river-bed be of a character enabling it to withstand the effects of the current, the dam itself has ample power of resistance. It has not been long under trial, as it was built in 1873; but it is thought to have undergone a sufficient test to prove it a reliable structure.

Our engraving presents a cross section of the dam, and one of the crib abutments. The builder is Mr. A. Garnsey of Sanford, Maine, who states that the stream has a hard gravelly bottom—not a ledge—and that rough stone, which it will be seen is the material chiefly used, is abundant in the vicinity. The length of the dam, across the stream, is 130 feet, and the extreme hight 10 feet, from the bottom of the up-stream mud-sill to the top of the cap-timber forming the crest of the dam.

In beginning the work of construction, the bed of the stream was dug down to the "hard-pan," and the two mud-

STONE APRON DAM.

sills were laid (the ends of which are shown in the cut) cross-wise of the stream, their size being 12 by 12 inches, and the distance between them 15 feet. Between these sills small stones were filled in, to a level with the top of the sills, as shown in the engraving; and upon the foundation consisting of the sills and stone filling the wall of which the dam is composed was laid. The kind of stone employed, and the manner in which it was laid, are so plainly shown as to require no explanation. Upon the up-stream mud-sill are erected the posts, against the upper and nearly perpendicular face of the dam (which requires to be as smooth and even as practicable) these posts being 6 by 6 inches, and placed at intervals of 2 feet. Upon the summit of the wall, and with its upper surface flush with the top of the posts, is a cap timber corresponding with the mud-sills, and of the same size, 12 by 12 inches.

To the upright posts the planking is secured, extending horizontally across the stream, 2-inch planks being used, and laid snugly together. Planks are also nailed, as will be seen, to the up-stream mud-sill, making them continuous from the base to the crest of the dam. Finally, a cap-piece consisting of a very thick plank is pinned or spiked on the top of the posts and the cap-timber previously mentioned, its upper surface being flush with the top of the planking, and the water being thus carried over the comb of the dam without tending to displace the 12-inch timber.

The special object of the builder of the dam in adopting this plan was, as stated by him, to avoid the expense of constructing an apron of timber, stone being very cheap and abundant. He suggests the laying of stones below the mud-sill on the down-stream side, taking care to place them together in such a manner that the water will not roll them out of place; and he adds that the further they are extended down-stream, the better. This feature is illustrated in the engraving, in which it will be seen that the base of the dam is continued down-stream for a considerable distance, large rocks being used, set squarely together, and so firm in their position that the current will pass over without disturbing them.

If the bed of the stream is clay, or of a sandy nat spiling should be driven down, against the upper sid the up-stream mud-sill; but the river bottom, in the of the dam here shown, was too hard to admit of spi being put in, even were it desirable.

The abutments, one of which is shown in our illustrat are cribs of the usual description, logs being laid up the interior filled with small stones, with which other s ble materials may be mixed, to make a compact mass prevent leakage, as described in former chapters.

CHAPTER XLI.

PILE AND FRAME DAM.

The description and illustration given in this chapter, like those in Chapter XXXVI, were elicited by the inquiry signed "D. P." in Leffel's Mechanical News for November, 1873, "how to put in a dam with 10 feet of quicksand at the bottom." The dam here shown was built about ten years since in Mobile County, Alabama, by Mr. Andrew McGregor, now Superintendent of Vaughan's Mills at Moss Point, Miss., who states that the stream on which it is located has a quicksand bottom of eight to twelve feet depth. The average depth of water in the river was two feet, the width being 68 feet, and the power afforded, with 10 feet 2 inches head, was applied in driving two 56-inch saws, each cutting an average of 15,000 feet of lumber per day.

The first step in the construction of the dam was to turn the stream with a bay-dam of brush and dirt. A trench was then cut in each bank 16 feet long and 4 feet wide, down to the water-level of the river. Six rows of round piling were then driven, extending across the stream, the piles consisting of pine poles 18 feet long and 10 inches in diameter. The piles in each row were placed 5 feet apart from center to center, and the distance between the rows, up and down stream, was 6 feet from center to center. In the first or up-stream row there were 20 piles, making the total length of this row about 95 feet, extending at each end about 12 feet into the trench, above-mentioned. The other five rows each comprised 14 piles, forming, when all were driven, a crib or pen about 68 feet long and 30 feet wide, up and down stream. The front row of 20 piles, it should be further stated, was continued at each end, an addition being made at right angles with the dam, and extending 12 feet up the stream, along the bank, thus forming a rectangular wing at each extremity of the dam.

Tenons were then cut on the head of each pile, down to the water level, and cap-sills framed on each row across the stream, the sills being timbers 12 by 12 inches square. Upon the cap-sills were gained stream-sills consisting of timbers 10 by 12 inches, which were let in 4 inches on the cap-sills. On the front row a filling was then put in, between the stream-sills, of pieces of timber 8 by 12 inches, and 4 feet 2 inches long, making a level surface on this row. Flat piles 3 by 12 inches, 18 feet long, were then driven on the up-stream side of the front row of round piles, making close joints, and extending, as already stated, a distance of about 95 feet across the stream and into the trench at each bank. This flat piling was spiked fast to the cap-sill, and cut off on a level with the top of it. Flat piling, 2 by 12 inches and 18 feet long, was also driven along each side row of round piles, and across the lower end, or down-stream row of piles, thus completely enclosing the crib or pen formed by the round piling as above stated. The corners were made secure by driving the piling double at those points and lapping joints. The surface of the crib was filled in or floored with sheeting 2 by 12 inches, nailed fast to the mud-sills and the joints made as tight as possible, to prevent any chance of washing by water running over the dam.

The rafters consist of timbers 15 feet long, and 10 by 12 inches square, framed upon each stream-sill, the foot of the rafter standing 4 inches back from the face line of the front mud-sill, so that the bottom plank on the face of the dam can be beveled to a tight joint on the stream-sills and the filling of short 8 by 12 timbers between them. The rafters have 10 feet 2 inches rise, that being the perpendicular hight of the dam from the stream-sills to the crest. Each rafter is provided with three braces of 10 by 12 timber, mortised into the stream-sill with 3 by 12 inch tenons. The rafters and braces are framed with a view to throwing the strain

PILE AND FRAME DAM.

downward and backward, as will be apparent on an inspection of the engraving.

The face of the dam is constructed as follows: for the lower portion, reaching one-third of the way up, five rows of planks 3 inches thick and one foot wide were put on, the lower plank being beveled at its bottom edge to a tight joint with the stream-sills and filling as already specified; then five courses of 2-inch planks, same width, and with close joints; and for the last or upper third of the face, five courses of 1½-inch planks, the upper three courses being laid without nailing, so that they could be taken off in case of high water. The 3-inch planks of the first five courses were fastened with 6-inch spikes; the 2-inch planks with 4-inch spikes, and the first two courses of 1½-inch planks with twelve-penny nails.

Against the face of the dam a filling was put in of sand and clay, extending up the slope a distance of five feet. This filling is not shown in the illustration, being omitted in order to present more clearly the construction of the dam itself.

The manner in which the wings or abutments of the dam are built requires some specific explanation. The front row of piling, as already stated, is so extended as to form a wing at each end of the dam, projecting 12 feet into the bank and turning 12 feet up-stream; and cap-sills were framed on top of these piles in the same manner as the others. In each wing were then placed five posts, 11 feet long, of 10 by 12 timber, framed on the cap-sills, making the abutment wings 12 feet high. Flat piling 2 by 12 inches and 22 feet long was then driven down, 10 feet into the ground, and spiked to the cap-sills, 24 piles being required for each wing. A filling of sand and clay was then put in and packed close. When thus finished the wings are not seen, and in order to show them the filling is omitted in the engraving. The object in turning the wings up-stream is to prevent the possibility of the water working its way around the abutments. A similar arrangement, it will be observed, protects the bank at the end of the dam from being washed out.

Mr. McGregor states in regard to the preliminary part of the work:

"It requires both skill and patience to get piling through quicksand. I used two 600 lb. hammers, and derricks 26 feet high, raised by two mules with snatch-blocks and tackle. Bolted on the derrick in front I had two strong bars of iron; the first, one foot above the sill, and the other six feet above, to form guides for piling. After placing a pile, I worked it back and forth to settle and line it; then made it secure, and let the driver fall four or five feet, letting the hammer rest on the pile for about one minute, and shaking gently after each fall. By this means I got each pile home, *settling* them through the sand rather than driving, and making a perfect job much quicker than trying to force them through by hard driving. I have followed mill wrighting for twenty years, and speak from experience."

As the mill-foundation accompanying this dam is of somewhat peculiar construction, we give a brief abstract of Mr. McGregor's account of the manner in which it was built. It is, in fact, simply a down-stream continuation of the foundation of the dam. Nine rows of round piles were driven, 5 feet apart across the stream, and 10 feet apart up and down the stream, the first row being 10 feet from the lower end of the crib or dam. Tenons were cut on the head of each pile, and mud-sills mortised on, thus constituting an addition of 90 feet to the foundation-crib of the dam. Timbers extending 90 feet, and 10 by 12 inches in size, were then framed on the 4th, 8th and 12th stream-sills of the dam, continuing those sills down-stream and making their total length 126 feet, the width of the addition, cross-wise of the stream, being 38 feet from out to out. This foundation was floored with 2 by 12-inch plank, nailed to the mud-sills. Posts were mortised and framed in, 10 feet apart, 14 feet between joints, for 110 feet, or up to the head of the rafters of the dam. Two posts were then framed on the face of the 4th, 8th and 12th rafters, and stringers put on, 126 feet long, making the mill floor 36 feet wide in the clear and 126 feet long. The dam being 68 feet in length across the stream, and the mill 38 feet from out to out, a space was left, it will be seen, of 15 feet on each side of the mill. This space was occupied by the water wheels, which were of the description known as "breast wheels"—one on each side of the mill, 15 feet wide and 13 feet 6 inches in diameter, and giving 20 feet distance from the head of dam to face of wheel. The flumes were 15 feet wide, 20 feet long, and 2 feet deep, and built in the following manner: On each side of the mill, two of the top-boards on the face of the dam were cut off, flush with the mill posts and the abutment posts. Three posts were framed in on each of the outside stream-sills. The posts next the wheels were 10 by 18 inches, and cut out to a radius of 13 feet 10 inches, so as to admit 1¼ inch flooring, nailed on to form a shroud, to increase the length of time for which the water is retained in the buckets of the wheel. The outsides of the flumes were nailed to three short posts set for that purpose, and the inner sides to the posts of the mill.

The builder of this dam and mill structure found that by the plan above described a very material saving was made in cost of frame; and he also asserts that the work was much firmer than if the two were separately built—the frame adding strength to the dam and foundation, making it as solid as a rock for the mill.

In our present engraving, as in many preceding ones, the dam is shown as if cut in two lengthwise of the stream, giving a cross-section of the structure, and exhibiting nearly every part very clearly; the river-bed being also shown as if dug out against a part of the up-stream face of the dam, in order to show the position of the flat-piling with which that side is protected, the upper part of the piling only being visible.

CHAPTER XLII.

PILE AND BRUSH DAM.

In the present chapter we describe a dam which, though embodying principles of construction that have already appeared in different combinations in preceding chapters, constitutes as a whole an arrangement of materials sufficiently novel to justify a somewhat minute account. The dam here illustrated is located at Mormontown, Taylor County, Iowa, and was built in the summer of 1871, by Messrs. Thos. King & Co. The stream on which it is situated is a small one, being but about 60 feet wide, with a sandy bottom and sand or clay banks. The nearest point at which stone can be obtained in sufficient quantity to build a dam is 15 miles distant; and as the cost of such material, delivered at the spot, would have been too great for the economy it was desired to practice, resort was had to other methods of construction. Two or three dams had been washed out (in what manner they were built we are not informed) at the point where this one now stands; but this has up to the present time withstood all the floods which have occurred, and has shown no signs of giving way. Brush and prairie sod were the substitute adopted in place of stone; not as being a preferable material, but on account of the inconvenience of obtaining the latter.

In beginning the work, piles were driven in a row across the river, two feet apart, the row making an angle instead of a direct line, the middle being some six feet farther up the stream than the ends at the banks. A row of piles was also driven along each bank, extending down the stream for sixty feet and turning outward at the lower end. The

PILE AND BRUSH DAM.

piles used were mostly of burr oak, about 12 inches in diameter, trimmed and sharpened and driven top in the ground, and squared at the top or upper end for tenoning as hereafter described. Short piles, sawed 6 inches wide throughout, and 2½ inches thick at one end, tapering to ¾ inch thick at the other end, were then driven all around on the inside of the main piling. Above the sawed piles or sheeting, planks were put on, 2 inches in thickness, and the main piling thus sided up.

Along the entire length of the piling, as will be seen in the engraving, both at the banks and across the stream, a cap-timber is put on. This timber is 8 by 8 inches, and to receive it a tenon 3 by 8 inches is made on the top of each of the piles.

Outside of the piling, and between it and the banks, below the dam, a filling of prairie sod was put in, this with the piling itself forming the abutment.

In constructing the dam, or what might be called the apron, the first step requisite was to lay the lower course of the three shown in the illustration. The work was therefore begun, at a point below the lower piles in the stream, by laying small trees or saplings, without trimming, placing them lengthwise, with the tops up-stream, and as thickly as appeared necessary to render them compact and substantial, the layer extending from bank to bank. This having been completed, a second layer was put on, six or eight feet farther up-stream; and this was succeeded by a third layer, falling back six or eight feet as before, and carrying the work so far up-stream that the extremities of the saplings reached above the piles driven across the stream. The brush was worked between the piles, and on their up-stream side the filling was put in to the hight intended for the dam, 8½ feet. For this filling, brush and prairie sod were used, sloping up-stream as shown.

The piles along the banks, forming the abutments of the dam, are some five feet higher than those in the dam itself, the effect of this elevation of the abutments being that when the water is high enough to run over them, there is no fall over the dam, but a comparatively level flow, and consequently no danger of washing below the abutments. The piles in this dam are driven into the sand from 12 to 15 feet. The water is admitted into the race at a point some 30 feet above the dam.

It will be seen that comparatively little material of an expensive character is used in the construction of this dam, and that a great part of the work could be done without the employment of skilled labor, the framing of the cap-timbers on the piles being the chief exception. The manner in which this dam was built renders it difficult to give a precise statement of its cost, Messrs. King & Co. doing most of the work themselves, with their own teams, and employing such assistance as was necessary by the day or month; to which they add, "we also had some help from our neighbors." They however estimate that the dam would cost, if all the material were to be bought and the labor hired on the usual terms, perhaps $2,000. Messrs. King & Co. are using a James Leffel water wheel to drive their grist-mill, and state that its performance exceeds their expectations.

It should be here remarked that by an inadvertence the planking on the further side of the stream, secured to the piles on their bank side, is made to appear in the cut as if put on up and down; whereas it should be in a horizontal position, as shown on the nearer bank, where the earth is represented as if dug away in order to give a view of the planking.

CHAPTER XLIII.

LOG AND PLANK DAM.

As regards cheapness of construction, with at the same time all necessary elements of durability and strength, a very satisfactory result was reached in the case of the dam illustrated in the present chapter; but it is also to be observed that the circumstances were highly favorable to economy, the cost of a large part of the material being absolutely nothing, aside from the expense of transportation; and this was not a formidable item, as everything required was procurable in the immediate vicinity. This dam was built in Lycoming County, Pennsylvania, in 1848, and has therefore seen sufficient service to entitle the plan of its construction to considerable confidence, as it still remains in good condition. The stream at this point has a gravel bottom; the west bank is sand and the east bank a sandy loam, and their average hight is about 4 feet above the water line on the up-stream side of the dam. The total length of the dam is 75 feet, and its hight, from the down-stream water line to comb of dam, 8 feet. Round timbers are used throughout (with the exception of parts of the abutments, as hereafter specified) and the planking is hemlock joists 3 inches thick.

For the foundation, five sills were first laid across the stream, their diameter being about 15 inches. Upon these, eight sills were laid, lengthwise of the stream, of the same size as the foundation-sills; and between these stream-sills shorter timbers were placed, parallel with them, laid snugly together, and with their down-stream ends flush with those of the stream-sills, but extending up-stream only as far as the large center log of the dam, which rests on the stream-sills and runs across the stream. The ends of this center log, and of three similar logs in a vertical line above it, are conspicuously shown in the cut. The purpose of the builder in allowing these timbers, intermediate between the stream-sills, to extend only to the center of the dam, instead of running them quite through to the up-stream mud-sill, was to let the interior filling, in the up-stream half of the dam, rest on the gravel bottom of the river; while below the center log, or in the down-stream half of the dam, the filling rests on the stream-sills and the shorter timbers lying compactly between them, giving weight and a corresponding degree of stability to the dam. Furthermore, the projection of the stream-sills and the short timbers in front of the dam, constitutes, as will be seen in the engraving, a very substantial apron, so that the water, after coming over the slope of the dam itself, strikes these projecting timbers and runs off smoothly, avoiding any reaction or the formation of a bar below the dam.

The length of the eight stream-sills is 25 feet, and of the projection in front of the dam about six feet. The short timbers between the stream-sills, to reach to the center of the dam, must therefore be about 15 feet long. They are put down as solidly as possible and "spotted" upon the foundation-sills on which they rest.

The next step was to lay the first and lowest of the four center-logs, already mentioned. For this purpose a timber as large as could be conveniently obtained was employed, 15 inches or more in diameter, the object being to give the dam as rapid an elevation as possible, a less number of ties being thus required. Other timbers of somewhat smaller size were laid parallel with it, as shown, and round ties or rafters placed on them, the upper end of each resting on the center log, to which they were notched and pinned, the lower end of the down-stream rafter pinned to the log resting on the stream-sills, and that of the up-stream rafter to the up-stream mud-sill. This end of the up-stream tie is therefore not seen in the illustration, being concealed by the stream-sill behind which it passes.

Another center log was then put on, directly above the

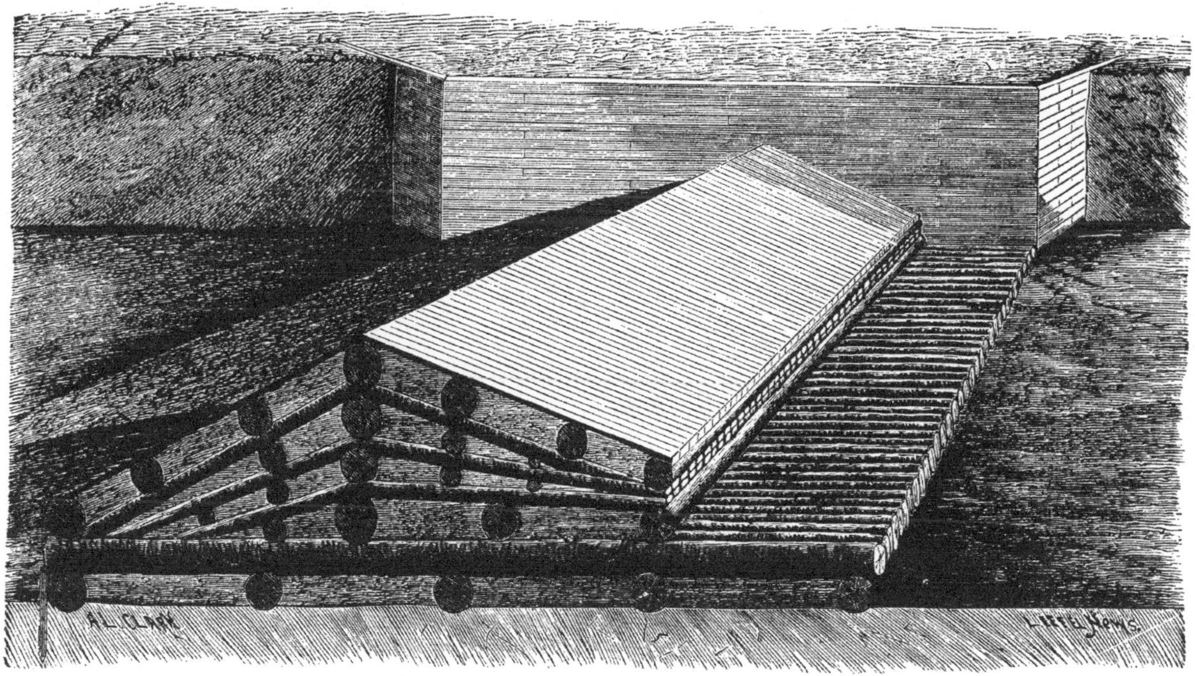

LOG AND PLANK DAM.

first, and ties notched and pinned on in the same manner as before, their lower ends being secured to the same timbers as the ties already laid. Then came a third center log, and a third tier of rafters or ties, notched and pinned and their lower ends resting respectively on the upper cross-sill at the down-stream face of the dam, and on the foundation-sill at the up-stream side. Between the successive tiers of rafters small timbers are introduced, parallel with the center logs and cross-sills, as shown in the engraving, their purpose being to give the rafters a solid bearing. Finally, the fourth or topmost of the center logs was put on, and on each slope of the dam three bearers parallel with it, and of about the same size, were laid as shown in the cut. These bearers were hewed on one side to receive the covering, both slopes of the dam being sheeted with 3-inch hemlock planks.

The whole interior of the dam was filled with stone and gravel; and on the up-stream side, from the bed of the stream up to the foot of the planking, a depth of about three feet, a filling was put in of round creek gravel, slate gravel being added on top of this until the entire filling extended about half way up the slope of the dam. We have indicated in our engraving a course of flat or sheet piling at the foot of the up-stream slope, although the builder of this dam informs us that no piling was used. We think it would be advantageous to introduce foot-piling against the up-stream mud-sill in the manner we have shown, securing it to the sill, as in this way the water would be prevented from working under the dam, in case the gravel filling should not entirely exclude it. The builder of this dam considers it best to extend the planking no farther down the slope, on the up-stream side, than the top of the sill or bearer resting on the stream-sill. By this method more space is given for the gravel filling than if the planking extended down to the river-bed, and he is of opinion that the gravel, having access to the mud and stream-sills, will effectually close the leaks and will in due time settle into a compact and impervious mass. If the planking is to run to the bed of the stream, he advises the digging of a puddle ditch or trench to render the foot of this slope perfectly secure against leakage.

One end of this dam abuts against the foundation of the mill; at the other end, which is the one shown in the engraving, is an abutment of the usual description. This is built of timbers hewed on three sides, and put up with the ties dove-tailed into the front timbers and notched into the back timbers. The ties and the back timbers are round. The face timbers of the abutment are 25 feet long, being equal to the width of the dam, and are notched, at the base of the abutment, upon the ends of the foundation-sills of the dam. At the ends of the abutment, the ties, hewed on three sides, are run into the bank at an obtuse instead of a right angle, the ties of the down-stream wing being about 12 feet and those of the up-stream wing 10 feet long. The interior of the abutment is filled with gravel and stone. Its hight is 6 feet above the crest of the dam, or 14 feet from the water line below the dam to the top of the abutment.

The heaviest of the timbers used in building the dam were generally, as already stated, about 15 inches in diameter, the lowest of the center logs being still larger than this. The ties or rafters running from the center up and down stream to the mud and cross sills were not more than 10 inches in diameter, on an average.

The bank on the side of the stream on which the abutment is located is not quite so high as represented in the engraving, being about two feet below the top of the abutment, or four feet higher than the water line on upstream side of dam.

The cost of this dam, as stated by the builder himself, will be considered extremely moderate. He says: "I put this dam up in 1848 for $153. I hired labor for $1.00 per day, and teams for $2.50, and filled the whole with stone up to the planking. The stone was handy; had to haul it but a short distance.

The plank cost, delivered, $5.00 per thousand feet, making	$ 28.00
Total labor and hauling	125.00
Total cost	$153.00

The timber was rough hemlock, cut on our own land, and

we did not consider it worth anything. When timber is plenty, I consider this as cheap and durable a dam as can be built."

It will be seen that the expense of constructing such a dam at the present time, or under circumstances less favorable as regards cheapness of materials and convenience of obtaining them, would be considerably greater than is indicated by the foregoing figures. This fact should be borne in mind in estimating the cost of any similar structure; but the full description we have given will enable any one interested to judge of the comparative cost of the work in a different locality and in a time of higher prices than were in vogue 26 years ago.

The mill for which this dam furnishes the motive power is situated, as above-mentioned, on the shore of the stream opposite to the abutment, the flume being directly between the mill-house and the dam, and a wing of the mill building extending out over the head-race. As the tail-race is carried along the side of the stream and in close proximity to the current, the builder says: "I run a small crib from the dam down the creek, built of small round timbers, averaging from 8 to 10 inches in diameter. The crib was 6 feet wide and about 4 feet high, and was filled with stone. The object was to keep the water, after it passed over the dam, from entering the tail-race. I started below the dam and sunk the tail-race down to gain head, making about 18 inches by that method."

CHAPTER XLIV.

FRAME DAM WITH SHEET PILING.

Upon either a mud, sand or gravel bottom, the description of dam presented in this chapter will be found, it is believed, a satisfactory one in all respects; and although not so economical in cost as some which have been previously illustrated, it still does not compare unfavorably with other plans, for the locality in which it was built. The best evidence of its merits is the fact that it has proved to be reliable where other methods of construction had signally failed. Preliminary to building this dam, (which is situated 2½ miles from Pittsburg Landing, Hardin Co., Tenn.,) a temporary dam of brush and dirt was thrown up to turn the water from the spot where the work was to be carried on. A ditch was then cut extending across the stream, from bank to bank, a distance of about 40 feet. A mud-sill 12 inches square was laid at the bottom of this ditch for its whole length; and in this sill mortices were cut at intervals of 4 to 6 feet to receive the posts. The posts are 12 inches square and cut to a length equal to the head of water it is intended the dam shall afford, which is in this case four feet. This is the rule adopted by the builder of this dam, his theory being that the pressure to be resisted by this part of the structure is the same, and requires the same length of timbers, as in the case of the superstructure. He however expresses the opinion that for a much higher head the length of the posts in the ditch need not be proportionally increased, and that, for example, posts 10 feet long would be sufficient if the dam were to give 15 or 20 feet head. In our own view, posts of much less length would answer equally well, as the lateral or

FRAME DAM WITH SHEET PILING.

sidewise pressure of earth is a very different thing from that of water. It need hardly be remarked that the depth of the ditch, the digging of which is the first item in the work, will be determined by the length it is intended to make the posts—the top of the post, when set, to be about one foot below the ordinary level of water in the stream.

The posts having been framed into the mud-sill, a cap-sill is put on top of them, reaching the whole length of the dam. A double row of sheet piling is then nailed on the up-stream side of the frame composed of the mud and cap sills and the connecting posts, the two courses of piling being so placed as to break joints. The piling may be of any suitable thickness, but should all be of the same width, in order that it may be snugly put on, one row covering the joints of the other. Another sill, 12 inches square and 40 feet long, is then laid across the stream, on the river-bed, at a point 20 feet down stream from the cap-sill on the posts. Stream-sills or sleepers of about the same size, and 20 feet long, are put on, their up-stream ends fitted to the cap-sill, and their other ends to the parallel down-stream sill. The stream-sills may be placed three feet apart. The ends which connect with the cap-sill are halved with it so that the stream-sill rests on the post and the cap-sill is let in upon the stream-sill as shown in the engraving. At its lower end, the stream-sill is laid on top of the cross-sill, and spiked or pinned to it. The flooring is then put down, for which purpose 2-inch planks may be used. In our illustration the flooring is shown running cross-wise of the stream; but if sleepers were laid across the stream-sills, the planks for the floor could then be put on lengthwise of the stream, which we think the preferable method.

Upon the platform thus constructed, 40 feet across by 20 feet up and down the stream, the posts, rafters and covering of the dam are placed. The posts are 12-inch timbers, of sufficient length to give the dam a perpendicular hight of 4 feet, and are set on the floor, the foot of each post being directly over a stream-sill, and firmly spiked down. Upon the top of this line of posts a 12-inch cap-timber is framed, and from this cap-timber the rafters, equal in number to the posts, extend to the up-stream edge of the floor, where their lower ends are secured with 12-inch spikes. For the slope here given to the dam, the rafters would require to be about 8 feet long. A shoulder is cut on the upper end of each to fit them to the cap-timber; and the rafters are then covered in the same manner as the apron or floor already described.

For the abutment, a ditch should be dug, and a mud-sill, posts and cap-sill put in, on the same plan as above indicated for the dam, with the same protection of sheet piling, in a double row, on the water side of the posts. A wall of hewed logs is then laid up, 40 feet long, its lower end being even with the down-stream extremity of the dam, and its upper end 20 feet up-stream from the ditch at the base of the dam. At each end, the abutment is turned at a right angle and carried back into the bank; and timbers may also be built up in the rear and ties inserted connecting the face and back of the abutment, forming a strong crib of the same general description as those represented in former chapters. The interior of this crib should be filled with earth, gravel or rock. The hight of the abutment here shown is about eight feet; and it should, in any given case, be carried up far enough to be a little above the highest stage of head-water, to prevent the stream from washing over or breaking around and injuring the banks. The cracks between the logs are stopped with planks to prevent leakage into the crib. In the engraving, planking is shown put on vertically, extending up to a level with the crest of the dam; and if the whole face of the abutment is covered in like manner, it will be all the more secure. Hewed logs are preferred to other lumber for erecting this abutment, as being more durable and equally satisfactory in other respects.

The chief peculiarity of this dam, and one which seems to recommend it for any suitable locality, is the manner in which the sheet piling at the base of the dam and abutments is secured, being fastened to sills at both the top and bottom of the piles, instead of driven down into the river-bed in the ordinary manner. Mr. John H. Macdonald of

Pyburn's Bluff, Tenn., who erected this dam in 1867, describes his previous experience in the use of piling as follows: "In 1866 I built this mill, and drove the spiling to a depth of 5 feet. I thought I had one of the best jobs in the world, and it was nice at the top. In June, 1867, my dam undermined, and everybody cried out 'Rat!' On examining it carefully, I found that it was the fault of the spiling, and when I ditched it out I found it in this shape: Stick your three fingers out straight—then draw your second finger forward—you can then see how my spiling was at the lower end. Now I suppose this is the case with all drove spiling; and no wonder the rats get so much abuse. Suppose a rat cuts a hole in the drove spiling, and the water commences running through; there is nothing at the lower end to hold, and the spiling begins to give way, one by one, till all is gone or ruined. The advantage my plan has over drove spiling is this: suppose a rat does cut a hole through the spiling, all the water can do is to pour through the hole. It cannot affect the dam, for the spiling is fast at the bottom and top. It must move the whole of the dam and mill and abutment at the same time. I feel assured that parties who build their dam and abutments in the manner I have described will save money in the long run, and the rats will go in peace."

Mr. Macdonald also states that at the point where he built this dam, four mill-seats had been undermined and ruined and three owners broken up; but the dam here described has now been standing seven years and is still sound. It is our impression that the flooring or apron planks of Mr. Macdonald's dam run lengthwise of the stream instead of across it, resting on sleepers transverse to the stream-sills as suggested in the foregoing article; but our engraving was made from data in which the transverse sleepers were not included, and the flooring was therefore represented resting directly on the stream-sills and cross-wise of the river.

CHAPTER XLV.

DOUBLE CRIB DAM—TRESTLE DAM.

A method of constructing a dam which may interest our readers by the somewhat novel features it possesses is presented by a structure of this kind on the Dakota or James River in Dakota Territory, erected by Mr. D. Shearer. The bed of the river is a mixture of clay and sand, and is not very solid. The banks are of clay, quite firm and reliable, 14 feet high on one side, while on the other side is a bluff some 75 feet high. The distance between them is about 100 feet. The dam consists of what may be called a double crib, with an intermediate filling, an up-stream filling or slope, and a double apron of stone on the down-stream side.

In the first place, a log crib was built, straight across the stream, 8 or 9 feet high, and of sufficient width to give it the necessary stability, say 8 feet. This crib was filled with rock and clay, and 12 or 14 feet farther up-stream another crib was built, similar to it but only 4 feet high, and of proportionate width. Stringers or ties were put in, extending between the cribs and holding them firmly together, and the space between them was filled with rock and clay. Above the up-stream crib, also, a filling was put in of the same materials, these being close at hand and costing little or nothing.

The apron is terraced, so to speak, consisting in other words of two long steps, giving it the form of the log aprons illustrated in previous chapters, where the successive courses of logs fall back as they rise, giving the water a gradual descent. The apron in this case is built of rock, the part next to the front or down-stream crib of the dam being 4 feet high and extending down-stream 14 feet. Below this is the second apron or step, also of rock, 2 feet high where it joins the first apron, and extending 12 feet down-stream, sloping at the same time to a hight of only one foot at its lower extremity. This gives a total extent of apron, lengthwise of the stream, of 26 feet, and the water is thus very smoothly and gradually carried away. The entire width of the base, from the up-stream extremity of the filling above the upper crib to the down-stream extremity of the stone apron, is about 60 feet, while the length across the stream, as above-mentioned, is 100 feet. With so broad a base, relatively to the hight, the latter being only 8 or 9 feet at the highest point, or surface of the down-stream crib, it will be seen that even with a strong current and a large volume of water, the dam is not likely to be carried off. The use of clay in the filling is the only feature which leads to any apprehension of injury by water working its way through; but the distance it must penetrate to do any material damage, and the use of rock in connection with it, render any disaster of a serious nature highly improbable. If a moderate proportion of sand or loam were also employed, so that any small inroad or gap which the water might produce would be speedily closed by the settling together of the materials themselves, there would be still less liability of trouble from this source.

The canal or race to convey the water to the mill is to be commenced at a point 10 or 15 rods above the dam. We should here remark that although we have in the foregoing account spoken of this dam as a structure actually completed and in service, it was in fact only in contemplation at the time our data concerning it were obtained; but we believe it has been built substantially upon the plan above described.

TRESTLE DAM.

For a locality in which it is intended to obtain but a comparatively low head, say not above 3 or 4 feet, a dam of very simple and cheap construction may be built, resembling somewhat the frame dams described in preceding articles, but of much less weight of timber and expense of framing. For a dam 3 feet high, trestles may be made of round timbers 6 or 7 inches in diameter and from 9 to 10 feet long, these constituting, with their covering of boards, the slope or upper surface of the dam. Near the upper end of each of these rafters three supports or legs are inserted, close together at the point where they enter the rafter, but spreading below in a direction parallel with the stream, thus giving them the requisite firmness against longitudinal pressure. The holes for insertion of the legs may be made with a 2-inch auger. The trestles thus constructed are set about 3 feet apart, with the legs down-stream, and the up-stream ends of the rafters settled thoroughly into the gravel or mud; or if the bottom is rock, the ends of the rafters should be beveled so as to fit upon it snugly. The trestles are set in a straight line across the stream, the rafters being in an even range so as to receive the covering in a regular and uniform manner. For the covering, inch boards are sufficient, secured to the rafters with nails. In putting them on, the work should be commenced at the top and continued toward the bottom of the slope. If the stream is a large one, a second course of boards may be put on, nailed on top of the first. This second course will run up and down, parallel with the rafters and the direction of the stream, the first course being necessarily laid crosswise of the stream. A filling of gravel, or of small stones, sand and loam is now put on the upper surface of the dam, and the work is finished. The strength of the fabric may be still further increased if it is underpinned with rock, and flat stones placed, broad side down, in front of the legs supporting the trestles. When thus strengthened, it is claimed that no liability of settling exists, and that the dam is absolutely secure from being washed out.

Mr. John F. Hazzen, of Sonestown, Penn., who furnishes us with particulars in regard to this dam, pronounces it, on the strength of his own experience, a cheap and good one. He states that he has used a dam of this kind at his mill for over twenty years. He also put in one of similar construction, five miles farther down on the same stream. This dam was but 20 inches high, and no difficulty was ever experienced with it. It resisted all the floods that ever passed over it, and when removed had to be torn out. The river has what is called a "rolling bottom," composed of gravel and sand, and is said to be a very hard stream on which to hold a large dam in place, in case of the water washing under it, and the dam consequently settling or tending to float off bodily.

CHAPTER XLVI.

LIGHT FRAME DAM.

The actual length of the dam illustrated herewith is 264 feet, and including the planking of the abutments amounts to about 300 feet; but in order not to reduce the scale of the engraving so low as to prevent the construction from being distinctly shown, the length of dam, as represented in the cut, is but about 100 feet, its hight to the water line being 7 or 8 feet. It is situated in Stephentown, Rensselaer County, N. Y., and was built by Mr. Eugene C. Goodrich in 1872, to run a circular saw and planing mill. The stream is a small and fluctuating one, and has a bottom of clay, gravel and mud. The object of the builder was to construct a tight dam, entirely of wood, to form a reservoir. The area of the reservoir thus obtained is about eight acres.

The foundation consists of three rows of large elm logs or mud-sills, laid crosswise of the stream, about 4 feet apart, each log from 12 to 24 feet long, as convenient; the logs of each row breaking joints with those of the next row. These are laid in trenches from 2 to 3 feet deep. Across these mud-sills are laid the cross-sills, lengthwise of the stream, 12 feet long, 9 inches square, and placed 6 feet apart. These sills are notched 1½ inch on the bottom, where they cross the three mud-sills, and pinned with 1½ inch pins. The mud-sills are spotted from 3 to 6 inches deep, and faced on the up-stream side. The cross-sills are also notched 1¼ inches deep on the top, a little up-stream from directly over each mud-sill, to receive the foot of the posts and the lower plate.

The posts, or braces, of which there are two courses, are

LIGHT FRAME DAM.

6 by 7 inches, the wide way being put lengthwise of the dam; and are pinned at the foot with 1¼ inch pins, without tenoning. There are three plates running lengthwise of the dam and supporting the rafters, each plate consisting of pieces 12 feet long, put in to break joints. The lower plate is 8 inches square, and rests directly on the cross-sills at the up-stream foot of the dam, in the 1¼ inch notches before-mentioned. The middle plate is 7 inches square, and rests on the short posts, which are about 3 feet long, a 1½ inch tenon being made on the upper end to hold the plate on, and a ¾ inch pin put in. The upper plate is 6 inches square, resting on the long posts, which are about 6 feet in length, and framed and pinned in the same manner as the short posts. The distance between the lower and middle plates is about 3½ feet, and between the middle and upper plates 4 feet. The posts are set at a right angle with the rafters, and at an angle of a little less than 45 degrees with the cross-sills. Braces 4 by 5 inches square are also put in between the middle and lower plates, over the cross-sills.

The rafters are 4 by 5 inches, and 3 feet apart between centers, every alternate one being over a post and cross-sill, and pinned with 1-inch pins. In the general view of the dam in our engraving, only half the rafters are shown, those occurring between the posts being omitted. The lower ends of the alternate rafters mentioned, which are those shown in our engraving, are notched over the up-stream ends of the cross-sills, and pinned, so as to bind all firmly together. The frame-work is all red oak. The planking is chestnut, 2 inches thick for about two-thirds of the hight from the bottom, and 1½ inch thick for the other third. Every plank is grooved, and tongues of oak ½ by 1¼ inch put in. This the builder considers a very important item in a tight dam. The planking in the abutment is also tongued, and even the spiling at the bottom. The planks are only 6 feet long, and are thoroughly pinned with ¾-inch oak pins.

At the up-stream ends of the cross-sills there are short oak posts driven down in the ground (of the same size as the rafters, 4 by 5 inches) and pinned at their upper ends to the cross-sills, to which are spiked 2-inch planks 12 feet long. To these the upper ends of the spiles are spiked, the planks also forming a tight joint at the angle between the spiling and planking. The spiles are of 1½ inch chestnut plank, tongued—though any kind of wood will answer, as being entirely under water at all times, it will last an indefinite period. The spiles are 4 feet long, more or less, according to the kind of bottom in which they are driven, and are sharpened on the flat sides and driven tight together. This is a slow and tedious process, but is of vital importance.

The ends or abutments are formed by perpendicular posts and planking, even with the foot of the dam, similar to the rafters and planking of the main dam, extending 12 feet into the bank at the west end and 24 feet at the east end. Below, or on the down-stream side of this upright part, there is a solid bank of dirt and brush filled in, about as high as the water line, some 10 feet thick, and very firm and compact. There is also a filling of mud, dirt and gravel on the water side of the abutment, and along the whole length of the dam, extending about 3 feet up the planking.

The apron is 60 feet long, and, as will be seen in the illustration, its upper edge is even with the upper plate of the dam. There are at present no flash-boards, but it is the intention of the owner to put them on the coming season. The apron is planked directly on the main posts of the dam, (which are supported by braces between the middle and upper plates) with 2-inch planks, so that the water will slide instead of falling over it. At the bottom are 12 feet planks 3 inches thick, placed lengthwise of the stream on two long mud-sills, their upper ends resting on the third or down-stream mud-sill of the main dam, and pinned with 1-inch pins. At the foot of the posts, across the 3-inch planks, is a 2½-inch oak plank to receive the force of the descending water and the shock of flood-wood coming over the apron. One-tenth of the capacity of the apron will carry the stream at ordinary stage, but it sometimes rises very rapidly and to a great height.

No spiling is considered necessary at the lower end of the apron.

In addition to the general view, we give a smaller cut showing a cross-section of the dam, in which the parts above described will be readily identified.

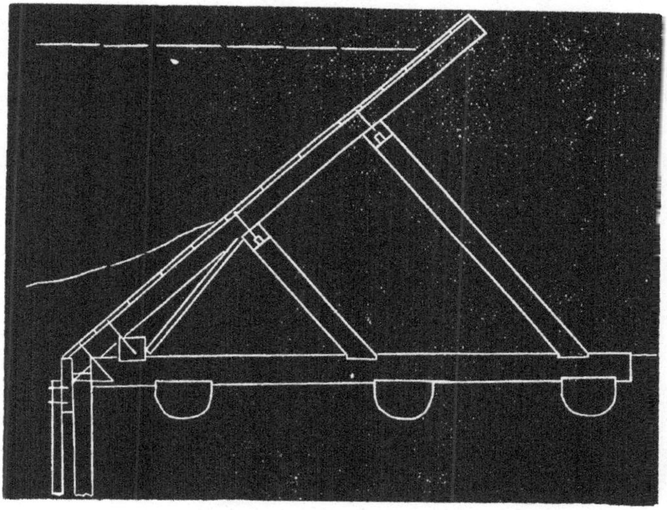

FIG. 2.

About 22,000 feet of lumber was used in the construction of this dam, valued at $550, and the cost of labor was about $950, making the total cost of the dam $1,500. The builder is confident that a cheaper dam of the length and size could not well be made, light timbers being used throughout.

The stream on which this dam is built takes its rise in Berry Pond, in Hancock, Berkshire County, Mass., the peculiar location of which renders it worthy of a passing notice. The pond, which is about three miles above the dam and covers an area of some 20 acres, is situated on the summit of a mountain, only a very small extent of ground being higher than the level of the water. The pond is at

an elevation of about one thousand feet above the su
of the reservoir formed by the dam.

CHAPTER XLVII.

DAM FOR ROCK AND SAND BOTTOM.

The stream on which the dam illustrated in this chapter is situated is known as the Little St. Francois, the location of the dam being 1½ miles north-west of Fredericktown, Madison Co., Mo. The course of the stream at this point is between high bluffs of porphyry, which seem to have been upheaved by some tremendous power beneath. The rocks are full of fissures, and the bed of the stream presents the appearance of large boulders, ten to fifteen feet in length and breadth, having been taken out at irregular intervals, and the cavities filled up with gravel and sand. The dam here referred to was built in 1869 by Mr. John T. Lee. He states that between the sloping solid rock bluff (shown in our engraving) and a large "hump" in the bed-rock in mid-stream, was a space filled with sand and gravel of unknown depth, in which he buried large white-oak logs, below the general level of the bed-rock, with the large ends down-stream, and extending about 8 feet below the face of the dam, on which to place the apron for this portion, no other part requiring any. Gains were cut in the round logs, and by chipping the bed-rock a place was made for the mud-sill as low down as possible. This mud-sill, running across the stream, was keyed in the gains in the logs and bolted to the bed-rock. The method adopted for bolting to the rock was to drill a hole from 6 to 10 inches deep, take an inch bolt and batter the end until it could just be pushed to the bottom of the hole, set it perpendicular and pour in melted lead or brimstone until the hole was full. The timber was then put down and fastened with a nut.

The dam is 110 feet long, 10 feet high from level of tail water, and 18 feet wide, and is built of sawn white-oak timber. The down-stream mud-sill is 10 by 12 inches; the cap-sill 8 by 10 inches; the upright posts 8 by 8 inches, and put 6 feet apart, mortised into the cap and mud-sills with short tenons and not pinned. There are also two up-stream mud-sills, the first put as low as possible, in a level position; and at intervals of 6 feet, timbers 6 by 8 inches, 18 feet long, are placed on the up-stream and down-stream mud-sills, serving as cross-ties, and bolted to the down-stream mud-sill with $\frac{3}{4}$ inch bolt and nut. The top up-stream mud-sill was next put down, and bolted down through the cross-tie to the lower sill, a nut being used wherever access could be had to it on the under side with a wrench. The face or down-stream side of the dam inclines up-stream about one foot from the perpendicular, and the cap-sill is bolted to the solid bluff. The rafters, which are 3 by 8 inches and 2 feet apart, were put on in the following manner: a gain was cut in the upper corner of the top up-stream mud-sill, in which to place the foot of the rafter in such a manner that in order to slip down-stream it must slide up an inclined plane. The cap-sill was also gained so that the rafter might have a bearing the full width of the sill. The rafter was gained at this end so as to give it a shoulder about $\frac{1}{2}$ inch deep on the down-stream side of the cap-sill, and extended about 10 inches beyond the cap-sill, so that the water might fall clear of the mud-sill. Spikes 12 inches long were made out of $\frac{1}{2}$-inch rod iron and driven with a sledge-hammer through the ends of the rafters (which were bored for the purpose) into the cap and lower sills. Under each rafter were put two braces, 4 by 6 inches, half the upper end of each brace being cut out to form a shoulder for the rafter, to which it was bolted with a $\frac{5}{8}$ inch bolt and nut. The lower end of each brace stands on the solid bed-rock, the brace leaning up-stream.

The first plank was then put on the lower ends of the rafters, its edge being beveled so as to make a true face with the top mud-sill. The spiling was then put in at the up-stream foot of the dam, in the following manner: oak

DAM FOR ROCK AND SAND BOTTOM.

planks 10 inches wide and 1 inch thick were sharpened at one end, wedge-shape, from one side only, driven down and drawn up again, and the battered places re-sharpened until the whole edge was of uniform shape. The plank or spile was then set and nailed to both sills and the covering plank, the beveled side of the spile being down-stream, or next the dam. The row of spiling was then doubled, the second row being of the same lumber, breaking joints with the first row, but having its beveled side up instead of down-stream. A filling of sand and gravel was put in, up to the top of the spiling on its upper side; and under the dam, against the two up-stream mud-sills, loose boulders were put in, extending up to and among the rafters for about one-third their length. The dam was then double-planked with inch boards, the first layer being of oak, the upper one of pine. The preference was given to pine as being less liable to warp in the sun. At the top of the dam, to finish it, a 2-inch oak plank was laid and well spiked on.

This dam has now been standing nearly five years, and has sustained several very heavy freshets, but at last accounts presented no appearance of yielding in any part, except the top plank just mentioned, which has been badly torn by logs and trees, these coming in innumerable quantities during the freshets, and rushing through the narrows with great rapidity.

The dam contains, in the aggregate, 12,000 feet of lumber, costing, at $15 per thousand, $180. The labor of two workmen, who built the dam in thirty days, is put down at $60; and the cost of nails, bolts, &c., is estimated at $25. The total cost of the dam, therefore, by the builder's figures, was $265.

OVERHUNG APRON DAM.

On pages 215 and 216 we have illustrated two dams of the "overhung apron" construction; and we now give a sectional view of another of the same general nature, but differing in some of the practical details sufficiently to warrant a brief description. The builder of this dam is

Mr. S. K. Cross of Burlington, Kansas, the location being on the Neosho River at that place. It was found necessary to remove a dam already in, which was of too light construction to be reliable, and erect this in its stead. The river-bed at this point is limestone. The bottom sills of the

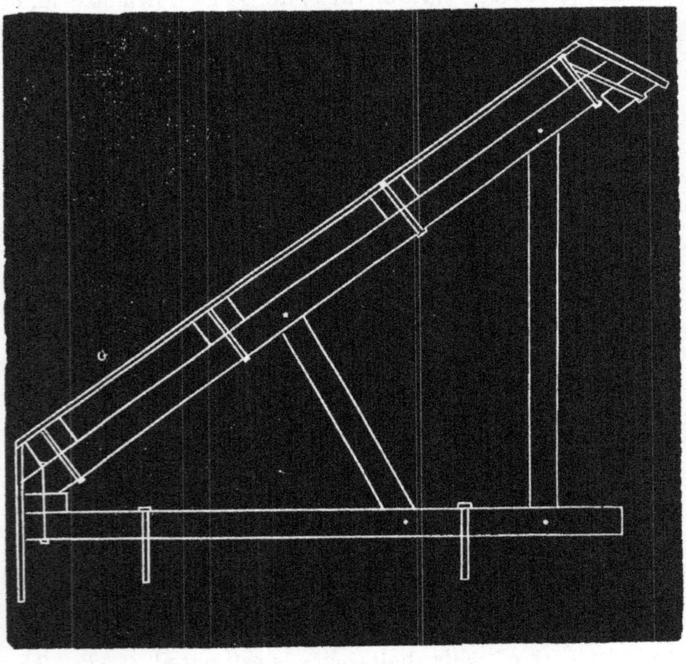

OVERHUNG APRON DAM.

dam are 12 feet long, 10 by 12 inches, and are fastened to the rock by two iron bolts 3 feet long and 1¼ inches in diameter, the holes being drilled so small that a heavy sledge is required to drive the iron pin into the rock. The ¾-inch bolt shown at the foot of the dam up-stream, passing through the rafter, cross-timber and bottom sill, is put through the bottom sill before the sill is laid, the head of the bolt being underneath and the nut on top. The cross-timber of which the end is here shown, resting on the bottom sill and supporting the foot of the rafters, is 10 inches thick, and need be hewed only on three sides. The same is the case with the lower cross-timber resting on the rafter. The two farther up the slope need only be hewed on two sides. All three are 10 inches thick with 10-inch face and are bolted to the rafters as shown, with ¾ inch bolts. At the upper end of the rafters the cross timber on their upper side is differently hewed, as indicated in the cut, to adapt it to the planking of the "shoot" or "over-hang." The timber bolted at this point on the under side of the rafter is 6 by 12 inches, and dressed as shown, its purpose being to give the projection or apron the necessary strength. The rafters themselves are 16 feet long, 10 by 12 inches, and the posts and braces supporting them 10 inches square.

The whole upper surface of the dam, including the projection, is covered with 3-inch plank, all the planks being dressed on the edge with a bevel of ¼ inch. Before dressing and using, they are allowed to lie until about half seasoned, this being in all such cases an excellent precaution against the injurious effects of either swelling or shrinking, one of which is likely to occur if either very dry or entirely green lumber is used. The cut indicates sheet piling at the up-stream foot of the dam, against the ends of the bottom sills and rafters and the side of the cross-timber between them, and making a close joint with the foot of the planking. On a rock bottom, of course, this piling could not well be put in, and Mr. Cross, we believe, merely planks this part, covering it tightly and making the joints very close.

The bents of this dam are placed 8 feet apart, between centers. The total length of the dam is 290 feet, and its hight 10 feet above the bottom sills. The Neosho rises to the hight of 24 feet in extreme high water. Mr. Cross states that he gains 5 feet in head-race and tail-race, and places the cost of his dam at $7 per foot, lumber being worth $30 per 1,000 feet. The overhanging apron delivers the water so far below the dam that it cannot wash up under the sills to affect the foundation; and the builder is confident that the dam will last till it rots out.

CHAPTER XLVIII.

RACE AND RESERVOIR EMBANKMENTS.

There is a radical difference in the conditions to be taken into account in constructing a dam for the purpose of raising the water in a river to a certain head and discharging the surplus, as compared with those which enter into the calculations for an embankment against standing water, or along the course of a deep and slowly moving body of water, as in a canal or race. In the case of a dam, there is but a comparatively narrow channel to be obstructed, and often a swift current is to be resisted; but on the other hand there are special means of fortifying the structure by securing it to the banks or to abutments, bolting its foundation to the river bed if it be of rock, building it with a curve or angle up-stream so as to give it the elements of strength pertaining to an arch, and in various other ways providing against the special dangers to which it is exposed. The peculiar sources of damage affecting the stability of a dam are the force of the current and the tendency of the water to wash it in passing over, (unless delivered through a chute or waste-way) or to react upon and undermine it when discharged by an overfall. In an embankment for a race or reservoir, on the other hand, while the advantages of banks or abutments to serve as an anchorage are not afforded, and the principle of the arch cannot be introduced, the sources of danger are at the same time fewer in number, and the problem is perhaps as much simplified as the means for its solution are reduced. The largest scale on which works of this character are undertaken is reached in the construction of levees and dykes; and in these but little difficulty would be encountered in securing the most ample strength and durability, were it not for the immense extent to which such works are necessarily carried, from which it results that the strictest economy in material and labor must be practised. A method of construction which would be thought but moderately expensive in a dam one hundred feet long would be absolutely ruinous in its cost if applied to a dyke or levee hundreds of miles in extent; and even when carried no farther than the embankment of a reservoir it would often require an impracticable outlay, as in this instance the burden falls on a single individual and is not borne by the whole community as in the case of a public work of this nature.

It follows that in the construction of embankments selection must be made from but few different kinds of material, and these of a cheap description—clay, sand and loam being the chief elements available for the purpose—and that the question of the plan of building resolves itself mainly into two points, the breadth of base and the angle of the slope. Of the three materials above-mentioned, we have in previous chapters given the preference strongly to sand and loam, in connection with gravel and rock, recommending the use of clay only to a very limited extent, and mixed with the other substances named. Our chief ground for taking this view is the tendency of clay to maintain its position when passages are worked through it, leaving these breaches open to be worn wider and wider, finally destroying the cohesion of the whole mass. Sand or loam, on the contrary, especially the former, will pour in and close a gap which may be accidentally made, and a leak may thus be stopped, in many cases, without any actual repair being required. This point we observe is recognized even by authorities which on general grounds advocate the use of clay. A commissioner of levees in one of the western States alludes to clay banks as being peculiarly liable to the attacks of craw-fish, which dig holes through them from the water side, and seize upon the small fish or water insects which pass in with the flow. He remarks that the clay embankments give the craw-fish every facility for his work, but that he cannot operate in sand, as the hole falls in as soon as made. He therefore advises that a wall of sand be carried up in a clay embankment, and urges this matter as one of great importance.

There is a marked distinction between the use of clay as a filling in crib-work, where a current is constantly seeking to force its way through, and the employment of the same material for a race or reservoir embankment, where its weight and solidity give it superior value. Except for the requirement of sand as a protection against craw-fish, it may be doubted whether a levee or embankment should not, in all cases where practicable, be built chiefly of clay. It is by far the heaviest of the three materials here under consideration, the weight of stiff clay being 135 lbs. per cubic foot, while that of loam is 124 lbs, and of light sand only 95 lbs. per cubic foot. No other single point in the construction of an embankment is of greater importance than that of weight, as this, with the proper breadth of base in proportion to hight, is the sole dependence for the solidity of the work. As to its impenetrability, which is another important item, it must be borne in mind that there is not the same liability to leakage in an embankment against comparatively still water as in a crib or filling against which a current is constantly directed, and over which the water frequently flows—to say nothing of the eddies, whirlpools, reacting currents and underwash which are perpetually threatening the safety of a dam. A clay embankment, therefore, which has its slopes not too steeply pitched, its material firmly packed down, and its face on the water side suitably protected, is as reliable a work as can be desired for the purpose it is to meet. It must be admitted that the use of sand for the whole body of the embankment, though it is frequently practised on account of the greater convenience of obtaining it, is open to serious objections. It is neither cohesive nor impervious; even the winds, as well as the waves and currents, have power to carry it before them, and if the water once succeeds in penetrating it in such a way that the leak does not instantly close from above, it will wear a channel with great rapidity, speedily enlarging to a crevasse. Loam is much less objectionable in this respect, as it is not only nearly thirty per cent. heavier, but is also much more cohesive and firm.

The slope which should be presented by the sides of an embankment is of course greatly dependent on the kind of material of which it is composed. The "standing angle" or "angle of repose" of the three substances above mentioned—that is, the steepness of slope which they will bear without sliding, when subjected to no other disturbing cause than that of their own gravity, is found to be, for sand, an angle of 30 degrees with the horizon; for firm loam, 36 to 45 degrees; and for clay, 55 degrees. In other words, supposing a pile to be made of each of these materials, coming to a point at the top, and having a perpendicular face on one side, as for instance if a bank of this kind be thrown up against a vertical wall, it is necessary that a bank of sand, in order not to slide or slip of its own weight, should have 1 foot 9 inches base for every foot of hight; for loam, about 1 foot 3 inches base is necessary to 1 foot of hight; while for clay 8 to 12 inches base to each foot of hight is sufficient. This being independent of all other forces or pressures than the weight of the material itself, the slope must of course be much more gradual, or in other words the base must be much wider in proportion to the hight, in an embankment against water, than is indicated by these figures; and the farther this excess of width in the base is carried beyond what the theoretical "angle of repose" would require, the safer will be the embankment. In practice, these mathematical proportions are in fact almost entirely disregarded, and the more sure although perhaps less scientific rule is adopted of giving the base so great a width as to provide against any possible tendency to slip; but the fact remains that for sand a much more gradual slope is necessary than for either loam or clay, and for clay a steeper slope may be permitted than for either of the others. In the embankments of the Welland in England a base of 70 feet is given for a hight of 8 feet, and on the Ouse, with 8 feet hight and a breadth of

10 feet on the crown, the base is 60 feet wide. In other localities the base is 5 or 6 feet to 1 foot of hight, divided, as to the two slopes, by giving on the water side 3 or 4 feet base, and on the land side 2 feet base to every foot of vertical elevation. On the sea-coast a still more gradual slope is given, being in many cases 4 or 5 feet to 1 foot to seaward, and 2 or 3 to 1 to landward. All these, it will be seen, are safe and substantial embankments, constructed rather with a view to permanent durability than to an immediate saving of expense; and this is a rule which can be confidently recommended in all works of this class, whether for public or private benefit.

The hight to which embankments should be carried is subject to variation according to circumstances; but the rule ordinarily adopted is to make the crown from 3 to 5 feet above high water mark. The breadth of crown is still more variable, ranging in different localities from 3 feet to 12 feet or more, the latter being an exceptional figure.

CHAPTER XLIX.

RACE AND RESERVOIR EMBANKMENTS—CONTINUED.

The necessity of adapting the slope of an embankment, especially on the water side, to the disturbing causes which will operate upon it, and the fact that it is better to err on the side of safety than otherwise, making the slope more gradual than is absolutely required rather than too steep, need not be further urged, having been fully demonstrated in the preceding chapter. The almost invariable tendency of earth to lose its stability in some measure when subjected to the action of water, and the fact that its impermeability cannot be depended on with certainty, render it necessary that the nature as well as the breadth of the foundation should be carefully attended to. There are, it is true, many cases in which embankments which were made by simply heaping up earth, without any special precautions to make it water-tight, have stood an indefinite length of time without exhibiting any defect; but in these cases the desired result has been secured by a lavish use of material without regard to economy, as in some localities in India where the cheapness of labor rendered it unnecessary to make such close calculations on this item as are elsewhere required. For a safe and permanent foundation, resort is sometimes had to a puddle wall with a "muck ditch" for its underlying support; the usefulness of this ditch being, however, strongly disputed, on the assumption that the natural surface of the ground is itself a more reliable foundation in many cases, the ditch being necessary only where the natural surface is loose and sandy, and by digging a ditch through it to a reasonable depth a firmer subsoil can be reached. Again, many embankments have been successfully constructed upon a foundation of brush, especially where the soil of the locality is light and unstable. This method is employed in building macadamized roads in marshy districts in England, and in dykes and embankments in Holland, Ireland and Canada. The most effectual manner of utilizing the brush is to put it down in two, three or four layers, their total thickness, when compressed by the weight of the overlying materials, to be from four to six feet; the brush to consist of branches of trees, as straight and tough as can be conveniently obtained, not too heavy, and of sufficient length to reach at each end within ten or twelve feet of the extremity of the slope of the embankment,—the brush, however, not lying directly across the foundation, but placed aslant, each successive layer crossing the preceding one after the manner of lattice work. The lowest layer may be pinned to the ground, and each of the succeeding layers to the one beneath it, by means of wooden forks; and when this method is adopted, the result is considered equally satisfactory with that obtained by the use of fascines, which are small bundles of brush tied up like a birch broom. In fact, the foundation can be more firmly knit and wrought together if simple brush is used than if fascines are employed.

There are also certain excellent methods of utilizing sand in works of this class, notwithstanding its weakness as a material for a simple embankment. A sand-bank with small brush and clay intermixed, and the slopes faced with a firm material, has been found extremely durable. The use of sand-piles, which is a somewhat novel substitute for piles of wood or iron, has been very successfully practised by English and French engineers. The method of setting such piles is very simple, and in a soft and deep soil, upon which a heavy embankment is to be placed, they are believed to afford the utmost attainable degree of security. Wooden piles are first driven, in rows along the middle and toward the sides of the foundation, and then withdrawn, the holes thus made being instantly filled with sand, which must be rammed down as firmly as possible as it is put in. Where the work is extensive enough to justify it, machinery may be employed by which a number of piles may be driven, withdrawn, and the holes filled simultaneously; and an iron cylinder, with internal screw-threads for sinking and raising, is sometimes used to facilitate the work. But either a light lift-ram or a heavy sledge will often answer the purpose sufficiently, and the piles may be driven one by one, each being separately finished before beginning on the next. The depth to which they should be sunk will depend, of course, upon the nature of the soil, but need seldom exceed six or seven feet. They should be 12 or 18 inches in diameter, and may be put from 6 to 10 feet apart. If the embankment is very heavy, it is well to have two or even three rows of piles in the central portion of the base, under what is to be the crown, and another row toward the outer edge of each slope; and the piles should be so arranged that the rows will "break joints," so to speak, the piles in one row not being directly opposite or in line with those in the next, but alternating like the spots on a chequer-board. It is somewhat surprising, in view of the unstable character of sand as a material to be used in bulk, unsupported by other elements, that it should be found so productive of strength when employed in this manner; but it is nevertheless a fact that its power of resistance in the form of piles, both in their lateral and transverse section, is superior to that obtained by almost any other form of construction, except those of the most expensive character. In an experiment by a corps of English engineers, nine piles, 4 feet 3 inches long and 8 inches in diameter were driven into a very soft soil, their distance apart being 16 inches between centers. To drive them, a weight of 200 lbs. was let fall from a hight of $3\frac{1}{2}$ feet, the driving being continued until the piles yielded only about $\frac{1}{4}$ inch at each stroke, after which they were settled about one-fifth of an inch farther by placing upon them a load of ten tons. They were then withdrawn, the holes filled with sand, and 16 more piles sunk in the same way, the whole occupying a total area of 36 square feet. Under a weight of about 1,000 lbs. there was only a settlement of one-twenty-fifth of an inch, and under 30 tons weight, after a month had elapsed, the total amount of settling was only three-fifths of an inch. On the whole, it may be safely stated that a foundation of sand-piles, especially if covered with layers of brush in the manner already described, will constitute, even in loose, light or marshy soil, a solid support for a heavy embankment, the slope being made sufficiently gradual to prevent any liability to slip, and the face protected against wash. It may also be remarked that where the soil is so firm that piling of any sort is not required, additional strength is often given to the foundation by spreading on the natural surface a thick and level coating of sand, to which is added, if the ground is very wet, one part of hydraulic lime to six parts of sand. This is too expensive a method for works which are to be carried to a great extent, but may be adopted with good results in the embanking of a race or reservoir of limited area.

The introduction of a puddle-wall is so often resorted to in the construction of embankments that it deserves a few explanatory remarks, and we also give in connection with this chapter two illustrations of the manner of applying this principle in embanking operations. A rude substitute for puddling is found in many embankments in which the earth has been rendered very firm and compact by the tread of the workmen employed in depositing it; but this is chiefly the case in Eastern countries, where the work is often done entirely by hand, and the bank is therefore well tramped as it is carried up, without any express outlay for the purpose. It is also true that in any low embankment of very gradual slope, and composed of firm, tenacious clay, the puddle-wall is unnecessary; but in the majority of cases it has been found most expedient, on the score of ultimate economy of material and stability of the work, to carry up an interior wall of the kind referred to, and also to arrange the other materials in relation to it so as to give the most secure protection to the base and interior of the embankment. In the two figures of our engraving, examples are given of the methods adopted by the best European engineers, the second figure showing some advantageous features not embraced in the first. It will be

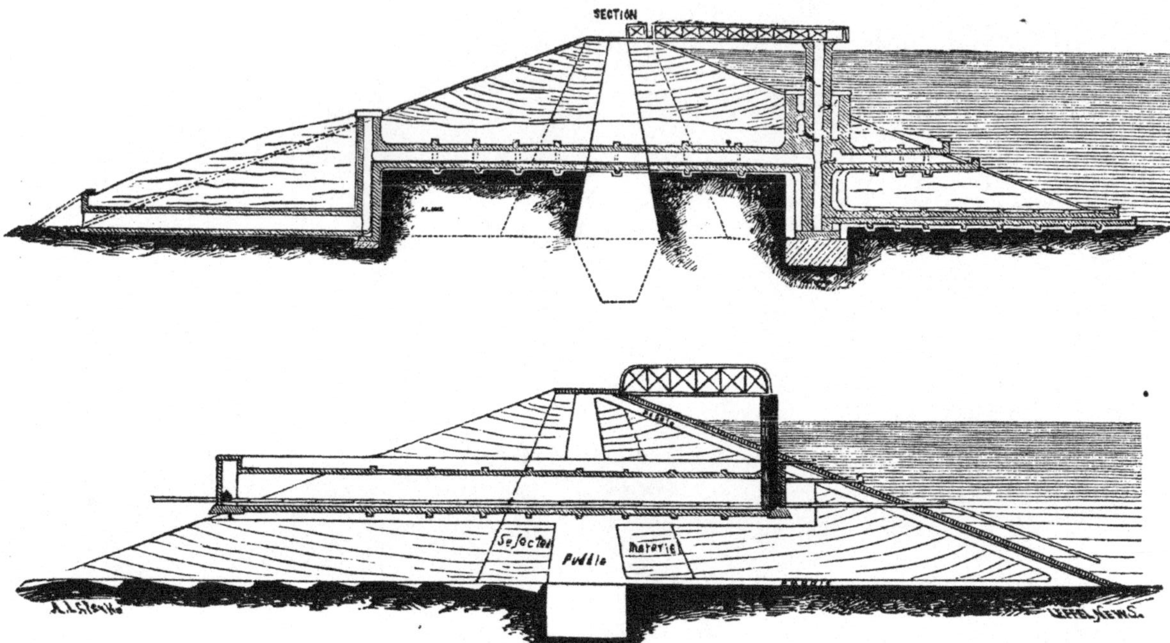

RESERVOIR EMBANKMENTS.

seen that the puddle-wall is in both cases carried down below the natural surface of the ground, there being in this instance a bed of rock or solid sub-soil which is thus reached, and the base of the puddle-wall being worked together and blended with it as thoroughly as possible, a nearly impenetrable barrier is thus presented to the water. Where there is no such solid underlying stratum, as has been already remarked, there is nothing to be gained by sinking the wall below the natural level of the ground; but in such cases it is well to loosen the surface of the ground to a depth of 6 inches or a foot, in order to unite with it the base of the puddle and in fact of the whole embankment as closely as possible; and the sod which may thus require to be taken off may afterward serve a highly useful purpose as a coating for the water side of the embankment. It is also worthy of note, that all grass, roots or other decomposable matter must be removed from the surface on which the foundation is to be made, whether it is subsequently used to sod the exterior slope or not.

In both the embankments we have illustrated, however, the ditch system at the base of the puddle-wall is adopted. In the first figure, this ditch has sloping sides, being much narrower at the bottom than at top; in the second, it has perpendicular sides. In both, the puddle-wall tapers from the ditch up to the crown of the embankment, being thickest where it has to withstand the greatest pressure. The pressure of water against a barrier thus erected against it is directly as its depth; and the shape of the barrier must of course conform more or less accurately to this law of pressure. The earth-works which we here illustrate are of the kind required where large bodies of water are to be enclosed, the lower figure representing, in fact, a cross-section of the embankment of the Biddeford (England) Water Works. For such cases, a breadth of 10 feet of puddle-wall at the level of the surface of the water is considered necessary for absolute safety; but in works of the kind where a less powerful pressure is to be sustained, the wall may be considerably reduced from that figure. The manner in which the remaining portions of the embankments are built is plainly indicated in the cuts. In the lower one especially, the arrangement of the materials used is a prominent feature, their relative position being such as to give the maximum degree of security. To this end, selected material, comprising the soundest and most tenacious and impermeable of the substances to compose the embankment, is placed next to the puddle-wall, on each side of it; although, as it is from the water side that the chief danger of a breach is to be apprehended, it would seem a still wiser method to place the best materials all on that side. Why the arrangement here shown is so generally adopted it would be difficult to state; but such appears to be the case. In the Biddeford embankment, it will be observed, the water slope is protected by a layer or coating of puddle extending from the foot of the slope to the crown of the embankment; and outside of this is a layer of peat. Another method, which is strongly recommended by many, is to cover the whole of the water slope with a layer of stone compactly laid by hand. Again, such embankments are often protected by growing grass thickly upon the top and sides,—an excellent method, for which Bermuda grass is said to be peculiarly adapted, as it grows very rapidly, and thrives both in sunshine and shade. A coating of loam should first be spread on to give the necessary fertility. This mode of strengtheuing an embankment is practised on many lines of railroad, and is equally adaptable to the protection of earth-works against water, one of its advantages being that the annual decay of the grass-tops affords each year a new though of course very light coating of compact matter on the slope, which in time adds greatly to its strength. In foreign countries thick ropes of twisted straw are often used for the covering of embankments, being pinned to the bank with forked sticks, the ropes lying so close together as to form a mat, which is still further strengthened by the grass or other vegetation working through and interlacing with it. In other cases, fascines, brush-wood, and sometimes large slabs of stone are laid upon the slopes of embankments to shield them from injury. A choice can readily be made from the different methods and materials we have enumerated, each builder suiting his work to the facilities at his command and the conditions he has to deal with. It should be mentioned that where the slope is protected by a layer of puddle, as shown in the second figure of our engraving, it is found beneficial to mix small stones or furnace cinders with the puddle to prevent the attacks of vermin. Fresh-water crabs have been known in some cases to take all the "pointing" from a wall of masonry, the mortar being highly useful to them as a material for the growth of their shells.

In both our illustrations is shown a system of pipes, culvert, valve-tower, &c., for providing the necessary outlet to the reservoir. The practice of conducting the discharge-pipe through or under the body of the embankment is strongly condemned by the best engineers, as it is thus rendered inaccessible in case a defect occurs and repairs are rendered necessary; and even if protected by a culvert, there is in such cases great danger of fracture resulting from the unequal settling of the embankment. Either with or without a culvert, therefore, there is incurred by this plan constant liability to accident, which may lead to serious damage to the embankment, and will certainly cause great inconvenience and loss in various ways. In the methods we have illustrated, the brick or stone culvert is located at a point one-half or two-thirds of the way up the embankment, and is made large enough to enable a man to enter it. A still safer plan is to carry the culvert around the end of the embankment, or to run a tunnel through the ground beneath and entirely clear of the embankment, there being in either of these cases no liability of injury to the culvert by the irregular settling of the earthwork. The plan shown in our first figure is substantially that designed by Mr. Rawlinson, an eminent civil engineer. The bottom of the culvert is in this case some 25 feet above the foot of the water slope of the embankment, the syphon pipe passing through the culvert. A shaft inside the embankment and connected with the horizontal culvert, contains the valves which conduct the water-supply from the reservoir, the valves, inlet pipes, &c., being so arranged and operated that the engineer has at all times full control of them, and all the parts are readily accessible for repair. In the cross-section of the Biddford embankment, also, similar arrangements for the discharge of water are shown, with the same communication of the culvert with the valve-tower, and the location in the latter of the inlet pipes at different hights, admitting of the drawing of water from the reservoir from points near the surface, where it is most likely to be pure. A float is sometimes attached to the end of a pipe inside the reservoir, this pipe moving up and down between guides as the water rises and falls, and having at its other end a flexible joint connecting it with the outlet pipe.

———:o:———

[Note Relative to Chapter XXXIV.—The builders of the dam at Osborn City, Kansas, described in Chapter XXXIV, in a letter dated April 15, 1874, give the following particulars, indicating the reliable character of their plan of construction: "We have had some high water this Spring. It has taken out two dams on the river, but ours is firm and all right. It is impossible to take it out, and we think it the best kind of dam that can be put in a stream that is not very wide. It has cost us a good deal, but the first cost is the cheapest in the long run. Parties putting in dams cannot do too much work on them. They should be completed in the start, and then you know you are all right." Mention is made in the same letter of a dam several miles farther up the stream, built of rock, logs and brush, but having no wing walls, or any protection for the banks—the result being that in the freshet above referred to, the water cut around the dam and nearly ruined the work.]

CHAPTER L.

HEAD-GATES FOR RACES AND CANALS.

The head-gate is so important a feature of the arrangements necessary for utilizing the power afforded by a dam, that it deserves at least a single chapter in a series like the present; and we therefore give herewith an illustration of a structure of this kind which is believed to possess important advantages. The engraving itself will be readily understood, without minute explanation, by all practical mill-wrights. Those of our readers who are familiar with the subject are well aware that an urgent need exists for a better plan of construction for head-gates than the old lift-gate, which has a troublesome peculiarity of being almost invariably out of order just when it is most needed. The majority of head-gates are hoisted and exposed to the warping influences of the sun and weather, so that in a short time it requires more patience than an average mortal possesses to shut one of them down; and in some instances they cannot be got down at all. The device shown in the engraving is intended to obviate this difficulty, and is claimed to effectually accomplish the object with but little more expense than is required by the old method.

The frame-work can all be put in to suit the location and the nature of the soil. In the engraving, sheet-piling is represented at E E E. If piling of this sort is required, it may be put down as deep as the builder may consider necessary. At G are shown the rack-bars, the object of which is to prevent trash and drift-wood from going in and down the race. A, B and C represent three head-gates in three different positions, hung on suitable axles or journals. The one at A is shown as it appears when closed; that at B, partially open, or in process of being shoved down to full open; and that at C is shown full open, with the pike-hook inserted in the stirrup, ready to pull up and close the gate.

The frame-work of the structure is made and planked up, outside and inside, in the same manner as an ordinary head-gate. It is then planked down from the top below the water line at both ends, and also planked over the top, trap-doors being provided, as shown, to afford easy access to the gates.

In the smaller figures, D and F, of our engraving, is shown more minutely the method of constructing the gate and framing in the journal boxes. The cap of the journal box can be keyed down on the journal as shown at F. The entire structure, including the gates and all other portions, can be made without nails or spikes.

The pike-pole, it may be here remarked, is an instrument which should be kept conveniently at hand about every water-mill, if for no other purpose than to drag heavy drift-wood out of the water.

When the trap-doors are kept closed there is no danger of the water freezing, even in the coldest weather, it being entirely protected from exposure. As the gates are generally open and therefore, of course, under water, they are not exposed to warping by the influence of the air and sun, but when needed are sure to fit and to be easily handled and effective. Many instances might be given in which defective head-gates have been the cause of the washing out and entire destruction of valuable mill property. Cases have also frequently occurred in which the head-gates, owing to their being either frozen up or warped to such an extent as to be unmanageable, could not be closed in time, and as a consequence a small wash or break in the head-race became the source of extensive damage, requiring a heavy outlay of money for its repair. It is easily seen in such cases, after the catastrophe has occurred, that true economy would have been consulted by providing in the first place a good head-gate which could have been easily reached and promptly closed, thus preventing any material

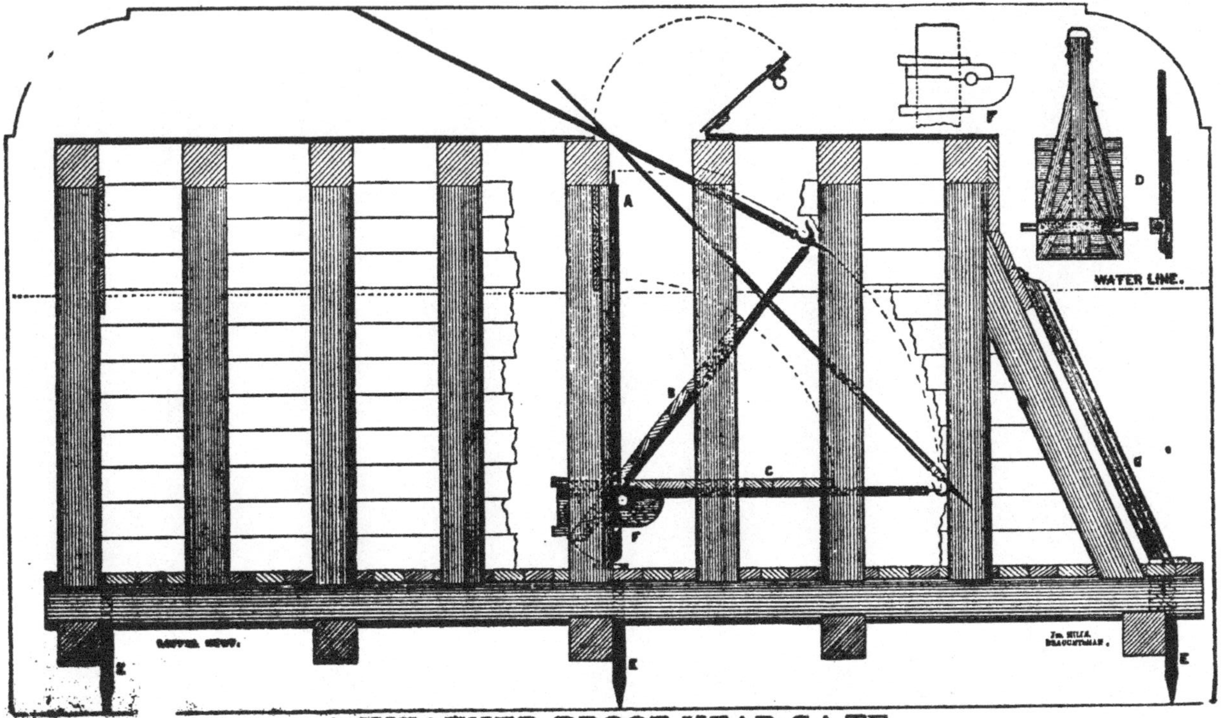

A WEATHER-PROOF HEAD-GATE.

injury. The teachings of such an experience are of course useful in guiding the subsequent operations of the owner; but in this as in many other matters, "an ounce of prevention is worth a pound of cure." The description and illustration which we have given of a head-gate adapted to meet just such emergencies will therefore be valuable to thousands of persons interested in the use of water-power, or who may hereafter engage in such enterprises.

By way of variety, we may here add a brief account of a device of this kind not so useful, in a practical sense, to the great body of mill-owners, because constructed on a scale and at an expense too great for any ordinary case. We refer to the guard-gate at the head of the Northern Canal at Lowell, Mass., as built some years ago, by James B. Francis, Esq. The canal had nearly perpendicular sides, an average width of 100 feet, and the water a depth of 15 feet. The gates were at the entrance from the Merrimac River, and not only regulated the supply of water to the canal, but protected both the canal and a portion of the city from the heavy freshets, often accompanied by the sudden breaking up of thick ice, which occurred in the spring. The greatest attainable strength in the walls and gates was required, especially for the reason that the entrance to the canal was in a nearly direct line with the current of the river, and the pressure brought to bear upon the works was correspondingly heavy. To meet these demands, the structure was so made as to admit the water to the canal through ten openings, each 8 feet wide and 15 feet deep, separated by hammered granite piers 2 feet thick, 15 feet high, and over 50 feet long, resting on a granite paving, beneath which was a natural foundation of hard slate rock. The openings were covered with granite lintels 2 feet thick, resting on the piers, wells being left for the gates, and the masonry being carried up above the lintels, under the gate-house, to a hight of 12 feet. The gates were made of white-oak timbers one foot thick, laid one above the other, and connected together by four bolts 2 inches in diameter, which passed down through the middle of the gate to a cast-iron plate forming the bottom of the gate.

The ends of the gate projected into grooves, and were faced with plates of cast-iron 7 inches wide and 1½ inch thick. These plates were made to slide on castings built into the masonry, the sliding surfaces being planed so as to make a nearly water-tight joint. The gates were raised and lowered by means of two vertical screws 17 feet long and 5 inches in diameter, with threads of one inch pitch, these screws passing through the hubs of a pair of gears, cut into female screws. These gears were operated from the water-wheel shaft by means of an intermediate arrangement of pulleys, belts, pinions and bevel gears, each gate being provided with two sets of pulleys and belts, one to raise and the other to lower it. The proportions of the gearing were such that the belt connecting the two pulleys had a motion 1,800 times as rapid as that imparted to the gate. The whole device was so successful in its operation that all the gates could be lifted at once, the process occupying, under ordinary circumstances, only about fifteen minutes.

CHAPTER LI.

GAUGING THE SUPPLY TO WATER WHEELS.

It is a fact well known to manufacturers, and users of water from hydraulics, canals or races, where several parties are drawing water from the same source of supply, that if this supply is not constantly greater than the full demand of all the water-wheels in use on the power, more or less contention will result among the users of the water. In such cases, A will declare that B's wheels are using as much water as his own, while they are only entitled to half as much. It is doubtful if there is a water power in the country, where no gauge is employed, which does not afford instances of some user of the water putting in wheels much larger than he had really the right of water to drive; and in such cases, if the gauge system were introduced, limiting each party to the precise amount of water he is entitled to, it would compel the withdrawal of wheels which operate wastefully in this respect, and the substitution of those of improved and modern construction. A device by which this accurate gauging of the water is accomplished has been submitted to us by Mr. Jonathan Mills, and from the drawings furnished by him the engraving has been executed which we present herewith. It shows only a front elevation of the plan, but will doubtless be readily understood by those interested in its object.

The method of the author of this device, as stated by him, would be to ascertain as follows, with a gauge of this description, what number of square inches of opening in the gauge it would require to furnish one hundred square inches of water, by measurement on a weir fixed in a tail-race, after leaving some well-known water wheel. For this purpose, he would first construct a gauge on the plan represented in our engraving, as near the required size as could be ascertained by calculation. The cut shows six gauges raised: let each gauge be supposed to represent an opening large enough to supply 100 inches of water, as measured in cubic feet on the weir after passing and doing its work on the water-wheel. The gauges are placed in the head race or flume above the wheel, a sufficient distance to allow a good body of water in the penstock. A float or buoy is then fixed in the water on the down-stream side of the gauge, with an arm extending up above the top of the flume. A pointer is fixed to the top end of the float-stem, resting against a board laid off in inches, from 1 to 10 inches inclusive—the board being stationary at the top of the flume. The water being on a level on both sides of the gauge, the pointer is set at 0; the wheel is then started, and the water measured as it passes the weir. Or the weir might previously have a gauge-mark set to a proper hight to allow just the required number of cubic feet to pass it when the limit of 600 square inches was used. The six gauges are then steadily closed, all being moved together until the float draws the pointer down to 10 inches. The gates are then opened on the water-wheel until the required hight on the weir in the tail-race is reached, the six gauges being at the same time raised just so as to keep the water at such a hight that the pointer shall not draw below 10 inches. As soon as the supply or gauge openings required to maintain the proper hight on the weir are determined, the next step is to shut down and get the exact opening of the six gauges. By this is established the size and form of each gauge on the power for each hundred square inches; and by this is determined the size of gauges to be used on the power, and at the same time the limit of draw that each user of water should be entitled to.

The question may be asked why 14 gauges are shown, when the limit in this case is made at only six. One object is to provide an arrangement by which A may rent of B, C, or D as many shares as the latter is not using. He may rent these shares for one day or night, or for any length of

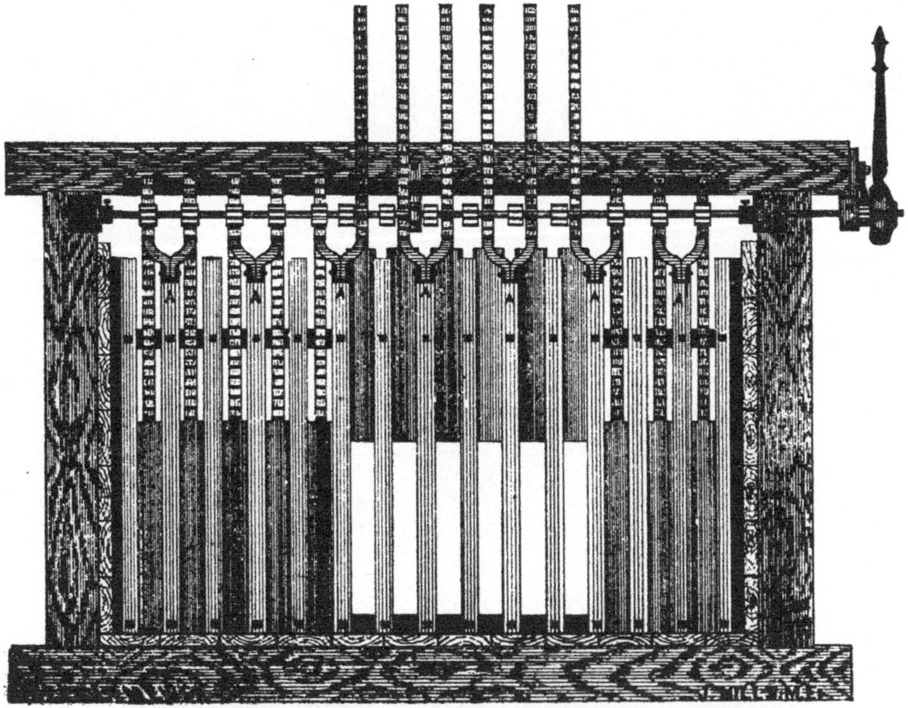

MILLS'S ADJUSTABLE WATER GAUGE.

time that may be agreed on. If the water is owned by a company, they should employ a person to go constantly from one flume to another and see that no pointer is drawn down below the limit agreed on by a majority of the users; and where a wheel is found drawing a pointer below the limit, he should have power to close the gates of the wheel until the pointer rises up to the limit, even if the works run by it are thereby stopped. His duty is to shut down the gauges of any party renting to another, and raise them for the party to whom the share is rented. Another object in allowing them to rise to any hight required is to permit all parties to use just what water they please when there is an abundance running to waste, which is frequently the case on all water-powers. The limit of draw should be established, by a majority vote of the users, on the first wheel, highest up on the race. They must also establish the amount of draw or fall that is caused in the main race, this depending altogether on the length and width of the race, and the quantity of water drawn out by different users along its course. The person employed to attend to the gauges should make his round each Sunday afternoon and set his gauge or draw-board up or down so that the pointer is at 0 on every draw-board on the race, and lock them fast. The reason for setting them on Sunday is that all the wheels are supposed to be shut down on Saturday night, and there is no other time when the head has attained its level from one end of the race to the other. By this means, the limit of draw is established every seven days; and it is far better to do one just thing on Sunday than to have every user of water doing injustice the remainder of the week.

If the draw in the race is established, between the first wheel on the race and the last one, to be three, six or nine inches, the last wheel will have a limit of draw three, six or nine inches more than the first. If the first one has a limit of 10 inches, and the draw of the race is established to be three inches, the last wheel has the limit of 13 inches to draw on; and so on in proportion all along the race. All these things must be controlled by a majority vote of the parties using the water. Once the gauges are established, and the users organized into a protection company, with their proper officers to keep the records of the power, it is but little trouble to meet once every week, or at the call of the superintendent, and establish the limit of draw, to be observed until it is necessary to fix another limit—the scarcity of water determining how often the limit will require to be established. In a time of very scanty supply of water it might be necessary to establish the draw at the very least limit,—which limit might be abandoned within twenty-four hours, in case of a sudden freshet, and the gauges thrown wide open for all parties to use all they could.

In reference to the engraving, it is proper to observe that there are several ways in which the apparatus might be constructed and reach the same results. The gauges are here represented as made of cast-iron, of a uniform width and length, with long arms extending up the proper distance, provided with cogs in which the pinion teeth and the ratchets or pawls A engage, to hold them to their places wherever raised. The letters A in the cut are just below the pawls, on the stationary guides between which the gauges slide up and down. These guides are bolted to a square frame of iron or wood, made of just the proper size to slip down between the sides of the flume or forebay. The guides are set the exact width apart to allow the gauges to slide between them. The gauges are raised and lowered by a shaft running in front of them, having pinions bored to fit loosely on a spline running the whole length of the shaft except at the bearings. One bearing is shown at the middle of the shaft. A flat cast-iron concentric piece might be made to extend clear across the gauges, on a level with the center of the pawls, locking all the pawls down at one operation, so that no one could disengage them to raise or lower the gauges except the superintendent in charge of the keys. The shaft is shown with a hub, with holes for the hand-lever, fitted to one end, the superintendent thus requiring to have but one lever, which will fit all the hubs on the power. If they are properly fitted up, it need be but very little trouble for him to unlock any gauge, slip in or out the pinion or pawl of the gauge and raise or lower it, lock it, and go on his rounds with the assurance that no one could change the gauges in his absence. Of course the pointer-boards should also be so fixed that they can all be locked and remain in plain view of any one passing by, without the necessity of entering any building to see them.

Every user of the water can thus see that his neighbor does not draw below his limit. The cost to each individual of putting in such a gauge would not be heavy, while it would insure to each his rights. Any arrangement that would be of utility would probably cost fully as much as this. Certainly, no effectual apparatus can be put in without cost; but there are many powers on which, if a fair and equitable gauge system were adopted, nearly every user of the water would be greatly benefited, and would be sincerely thankful for deliverance from the old freedom of using and destroying what he could not use with profit—only to wish it back when it was too late.

With this system of gauges established, it would make no difference to any user of the water what kind of wheel his neighbor was running, as the whole loss for want of power would fall on the party using an inferior wheel.

Mention should be made of one point omitted in the foregoing remarks, viz.: that the bottom of the gauges must all be set the same distance under the head-water—not taking the Sunday level for it, but making allowance for the draw in the race, and setting from that standard, which is fixed by a majority vote of the users on the power. When once the draw is established, it must be permanently marked and kept on record, never to be changed except by a two-thirds or three-fourths vote of the users.

It is doubtful whether any stationary gauge with a fixed opening will give as good satisfaction as the system of the float limit here described. There are times when an abundance of water is running to waste, with back water on the wheels, and every user would prefer an arrangement such that he could open his gauges and let the water run freely. The fixed gauge, it is true, is in use to a considerable extent, but numerous examples prove that it does not fully meet the demands of users of water-power.

CHAPTER LII.

WEIR DAM—MEASUREMENT OF WATER.

When a man has concluded to improve a water power, the first thing he should ascertain is the amount of head and fall he can secure. The next and most important step is to determine, accurately, the quantity of water that flows in the stream, as upon this will depend the amount of power and consequently the amount of work the stream is capable of performing. And as the improvement of water power is necessarily attended with expense, it is therefore important to one who contemplates building a mill or factory, that he should know exactly what amount of work he can depend upon the stream doing; as for the want of an accurate knowledge, or from erroneous supposition of the quantity of water in the stream, parties frequently construct mills and factories of a magnitude which, upon trial, is found to be entirely out of proportion to the power of the stream, and it fails to drive their machinery. Such being often the case, it is important, when practicable, to get some one well versed in hydraulics to measure the capacity of the stream. As this cannot always be done, we give herewith a few plain suggestions, by the aid of which any one can determine, approximately, the quantity of water in a stream.

The accompanying engraving represents a weir or dam across a small stream. Where it is convenient to use a single board as is shown in the cut, select one that is long enough to reach across the stream, resting in the banks at each end; cut a notch in the board sufficient in depth to pass all the water to be measured, and not more than two-thirds of the width of the stream in length. The bottom of notch B in the board A, should be beveled on the down

38

A
B
C
D
E
LEFFEL-NEWS

stream side; the ends of the notch should be also beveled on the same side, and within one-eighth of an inch of the upper side of the board, leaving the edge almost sharp. E is a stake driven in the bottom of the stream several feet above the board or dam, and should be driven down to the level of notch B, this level being easily found as the water is beginning to spill over the board. After the water has come to a stand and reached its greatest depth, a careful measurement can be made of the depth of the water over the top of stake E, as illustrated in the cut by the man with square and measure in his hand. Such measurement will give the true depth of water passing over the notch, since, if measured directly on the notch or board, the curvature of the water in passing would reduce the depth, giving an improper measure; although where accuracy is not required, such a method will give a fair estimate of the quantity of water. In all cases it is best to make the measurement over the stake. The line D is a level from the bottom of notch B to the top of stake E; while the dotted line C represents the top of water, and the distance between the *lines* or from the top of stake gives the *true depth* or spill over the weir. The lines have, in the cut, the appearance of running over the top of the board; but this is owing to the fact that they pass behind it, and for the purpose of illustration, the reader is supposed to look through the board and post. The surface of the water below the board should not be nearer notch B than ten inches, that the flow of water over the notch may not be impeded; neither should the nature of the channel above the board be such as to force or hurry the water to the board, but it should be of ample width and depth to allow the water to approach the notch and board steadily and quietly. If the water passes the channel rapidly, it will be forced over the notch, and a larger quantity will pass than if allowed to spill from a large body moving slowly.

When the depth of water over the stake E is known, the quantity of water passing can be easily calculated, by reference to the Weir Table on page 309. This table gives the number of cubic feet of water passing per minute over a weir for each inch in breadth, from one-sixteenth of an inch in depth to twenty-five inches depth. The figures 1, 2, 3, &c., in the first and last perpendicular columns, are the inches depth of water over weir, while the first or top horizontal column represents fractional parts of an inch, from one-sixteenth to sixteen-sixteenths. The body of table shows the cubic feet and decimal parts of a cubic foot, that will pass each minute, for each depth of weir from one-sixteenth to 25 inches as before stated, but each result is for but one inch in width; so for any particular number of inches breadth of weir, the result obtained in the table must be multiplied by the number of inches of breadth the weir may be. For example: suppose the notch or weir be 20 inches wide, and the water at stake E, 5½ inches deep; in the first or last column find the figure 5, follow the horizontal column, until the perpendicular column is reached containing ½ at the top. In the square where these two columns meet will be found (5.18) five and eighteen hundredths cubic feet; this is the quantity of water that will pass for each inch in width; but since the supposed weir was 20 inches wide, this result must be multiplied by twenty, which gives (103.60) one hundred and three and six-tenths cubic feet per minute. In this manner the water passing any width of weir, of any depth from one-sixteenth of an inch to 25 inches depth, can be easily calculated.

A very important matter in connection with the measurement of small streams is also the possibility of damming or holding the water, and using it a part of the time instead of constantly. If a given quantity of water was held for 12 hours, for the remaining twelve hours (if all was used in that time,) double the quantity would be available, and consequently double the power would be obtained for that length of time. The power is thus stored up to be used in less time, besides giving a better effect, since with a small quantity of water almost as much power is required to drive the necessary machinery without labor, as when driving it at labor.

TO MEASURE WATER MORE ACCURATELY.

It sometimes becomes necessary to vary the foregoing

Inches	Parts of an Inch.	1/16	1/8	3/16	1/4	5/16	3/8	7/16	1/2	9/16	5/8	11/16	3/4	13/16	7/8	15/16	Parts of an Inch.	Inches
		.006	.01	.03	.05	.07	.09	.11	.14	.17	.20	.23	.26	.30	.33	.36		
1	.40	.43	.47	.51	.55	.60	.65	.70	.74	.78	.83	.87	.93	.98	1.03	1.08	.40	1
2	1.14	1.19	1.24	1.30	1.36	1.41	1.47	1.52	1.59	1.65	1.71	1.77	1.83	1.89	1.96	2.02	1.14	2
3	2.09	2.16	2.23	2.29	2.36	2.43	2.50	2.57	2.63	2.71	2.78	2.85	2.92	2.99	3.07	3.14	2.09	3
4	3.22	3.29	3.37	3.44	3.52	3.60	3.68	3.75	3.83	3.91	3.99	4.07	4.16	4.24	4.32	4.41	3.22	4
5	4.50	4.58	4.67	4.75	4.84	4.92	5.01	5.10	5.18	5.27	5.36	5.45	5.54	5.63	5.72	5.81	4.50	5
6	5.90	6.00	6.09	6.18	6.28	6.37	6.47	6.56	6.65	6.75	6.85	6.95	7.05	7.15	7.25	7.35	5.90	6
7	7.44	7.54	7.64	7.74	7.84	7.94	8.05	8.15	8.25	8.35	8.45	8.55	8.66	8.76	8.86	8.97	7.44	7
8	9.10	9.20	9.31	9.42	9.52	9.63	9.74	9.85	9.96	10.07	10.18	10.29	10.40	10.51	10.62	10.73	9.10	8
9	10.86	10.97	11.08	11.19	11.31	11.42	11.54	11.65	11.77	11.88	12.00	12.12	12.23	12.35	12.47	12.59	10.86	9
10	12.71	12.83	13.95	13.07	13.19	13.31	13.43	13.55	13.67	13.80	13.93	14.04	14.16	14.30	14.42	14.55	12.71	10
11	14.67	14.79	14.92	15.05	15.18	15.30	15.43	15.56	15.67	15.81	15.96	16.08	16.20	16.34	16.46	16.59	14.67	11
12	16.73	16.86	16.99	17.12	17.26	17.39	17.52	17.65	17.78	17.91	18.05	18.18	18.32	18.45	18.58	18.72	16.73	12
13	18.87	19.01	19.14	19.28	19.42	19.55	19.69	19.83	19.97	20.10	20.24	20.38	20.52	20.66	20.80	20.94	18.87	13
14	21.09	21.23	21.37	21.48	21.65	21.79	21.94	22.08	22.22	22.35	22.51	22.65	22.79	22.94	23.08	23.23	21.09	14
15	23.38	23.53	23.67	23.82	23.97	24.11	24.26	24.41	24.56	24.71	24.86	25.01	25.16	25.31	25.46	25.61	23.38	15
16	25.76	25.91	26.06	26.21	26.36	26.51	26.66	26.81	26.97	27.12	27.27	27.43	27.58	27.73	27.89	28.04	25.76	16
17	28.20	28.35	28.51	28.66	28.82	28.98	29.14	29.29	29.45	29.60	29.76	29.92	30.08	30.23	30.39	30.55	28.20	17
18	30.70	30.86	31.02	31.18	31.34	31.50	31.66	31.81	31.98	32.15	32.31	32.47	32.63	32.80	32.96	33.12	30.70	18
19	33.29	33.45	33.61	33.78	33.94	34.11	34.27	34.44	34.60	34.77	34.94	35.10	35.27	35.44	35.60	35.77	33.29	19
20	35.94	36.10	36.27	34.46	36.60	37.87	36.94	37.11	37.28	37.45	37.62	37.79	37.96	38.14	38.31	38.48	35.94	20
21	38.65	38.82	39.00	39.17	39.34	39.52	39.69	39.86	40.04	40.21	40.39	40.56	40.73	40.91	41.09	41.26	38.65	21
22	41.43	41.60	41.78	41.96	42.13	42.31	42.49	42.67	42.84	43.02	43.20	43.38	43.56	43.74	43.92	44.10	41.43	22
23	44.28	44.46	44.64	44.82	45.00	45.18	45.38	45.53	45.71	45.90	46.08	46.26	46.43	46.63	46.81	47.00	44.28	23
24	47.18	47.36	47.55	47.72	47.91	48.09	48.28	48.46	48.65	48.83	49.02	49.20	49.39	49.58	49.76	49.93	47.18	24

method in certain particulars, when it is desired to ascertain with great exactness the quantity of water a stream furnishes or a wheel is using. On very small streams, or where wheels are competing, or where the useful effect of power of a wheel for the quantity of water is required with special precision, the arrangements for measuring should be more carefully prepared, and corrections made that are not taken into consideration in the foregoing description. The notch B should be made in a thin plate or sheet of iron, forming almost a sharp edge, and screwed fast to the board A, on its upper side, the requisite stiffness being thus afforded to the iron. It is important that a thin plate should be used, as a thick one retards the flow. The notch in the iron should be made sufficiently less in size than that in the board, both on the bottom and at each end, to enable the water to pass clear of the board at all points, its flow being thus unobstructed.

If the ordinary square and measure is used, the stake E should be driven so that the top will be precisely level with the edge of the iron lip or notch; but since the capillary attraction caused by placing a rule in the water and on the stake gives rise to some uncertainty in measuring by that means, it is best to use a hook gauge. In this case the stake E stands above the level of the water to any convenient hight, and is graduated with any degree of minuteness desired. The point on the stake on an exact level with the top of the notch may be fixed by means of a spirit level and straight edge. From this point to the commencement of the graduated scale, or zero, the distance is equal to the length of the gauge less the vertical length of the hook, so that when the water is even with the notch, the top of the gauge will be at 0 on the scale, the top of the hook being at the surface of the water. Then as the water rises, the gauge is held against the stake and carefully adjusted by sliding up until the hook comes, as before, exactly to the level of the surface water, when the top of the gauge will show on the scale the precise depth over the notch. The velocity of the water as it approaches the weir should be considered and carefully calculated, as its effect is to increase the discharge. The correction is made by adding to the actual depth obtained the amount of head water that would produce the velocity.

Then from this measure can be ascertained by the table the actual amount of water spilling; except that from another cause of less consequence, but of sufficient importance to engage our notice, there is also a correction to be made, which is for the contraction to which short weirs are subject at the two ends. Weirs of all lengths, especially if narrower than the channel, are liable to this deviation or narrowing of the stream or flow of water—not, however, in so great a proportion as short ones. Experiments of a thoroughly reliable character show that this condition of the spill of water operates at both ends, and reduces the effective length of weir in about the proportion of two-tenths of an inch for each inch in depth of the spill, or one inch for each five inches depth; that is, for a weir 80 inches wide and the spill 5 inches deep, the actual width to be calculated for will be 79 inches.

It rarely occurs that such exactness will be required for the measurement, as is here described, and for all ordinary and practical purposes, the preceding instructions will be found sufficiently accurate.

MEASUREMENT OF LARGE OPEN STREAMS.

As in many cases it is impossible to construct even a temporary waste-board or weir, the quantity of water that the stream can supply must be obtained by first ascertaining the mean velocity in feet per minute, and also the area of cross section of the stream in square feet; when the product of these two quantities will give the required quantity of water afforded by the stream. The velocity of such stream can be estimated by throwing floating bodies on the surface of near the same specific gravity as the water, and rating the time accurately, required in passing a given distance; it must be borne in mind, however, that the velocity is greatest in the center of the stream, and near the surface; and that it is less near the bottom and side. It is generally best to ascertain the velocity at the center, and from this estimate the mean velocity, which has been found by accurate and reliable experiments to be 83 per cent., or about four-fifths of the velocity of the surface. The cross-section may be estimated by measuring the depth of stream at a number of points, at equal distances apart (these points being in a line across the stream,) adding the depths together, and multiplying their sum by the distance apart in feet of any two points. This will give the result required in square feet of cross section, when the product of mean velocity in feet per minute and cross section in square feet, obtains the quantity of water that the stream affords in cubic feet per minute.

In place of the ordinary rectangular notch, Prof. Thomson, in a paper read at the British Association at Leeds, recommends the use of a triangular notch, especially for the measurement of very variable streams. The exception taken to the rectangular notch in such cases is that if it is made wide enough to allow the water to pass in time of flood, it must be so wide that for long periods, in moderately dry weather, the water flows so shallow over its crest that its indications cannot be relied on. Accordingly, it is proposed to make the triangular notch of a capacity sufficient to admit of its use in floods, yet adaptable by its own shape to the smallest flow. Prof. Thomson remarks that by the use of such a notch, quantities of running water from the smallest to the largest may be accurately gauged; for when the quantity flowing is very small, the flow is confined to a small space, admitting of precise measurement; and the space for the flow of water increases as the quantity to be measured increases, but still continues such as to admit of accurate gauging.

THE END.

IT IS NOW UNIVERSALLY ADMITTED THAT THE

"EUREKA"

Combined Smut AND Separating Machine

Is by far the most popular and reliable Wheat Cleaning Machine in the world, having a sale larger than all others combined, and used in every country on the globe, where there is wheat to be cleaned and mills to grind it. The superiority of this machine over all others has been conceded at the numerous Expositions and Agricultural Fairs in the United States and Canadas, by awarding the highest prizes given. The

Only GOLD MEDAL given by the Royal Agricultural Society of England in 1869

—WAS AWARDED THE "EUREKA." ALSO THE—

Medal for Progress at the Vienna Exposition in 1873

The highest that could be awarded, was unanimously given the Eureka over all the Wheat Cleaners exhibited by American, English and Continental manufacturers. Over 7,000 sold in the last six years, and the demand steadily increasing. Eight sizes and styles of machines built, equally well adapted to the smallest custom and largest merchant mills. For further particulars and full description, send for Illustrated Pamphlet and Price List, or address

Howes, Babcock & Co.,

(Sole Proprietors & Manuf'rs,) Silver Creek, Chautauqua Co., New York,

HEAVY HUSK PORTABLE MILLS.

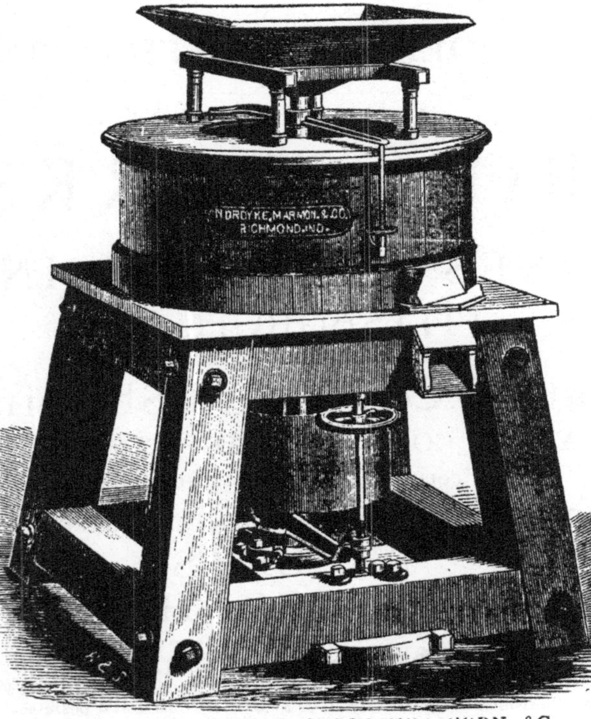

SUITABLE FOR WHEAT, MIDDLINGS, CORN, &C.

Four sizes are made—26, 30, 36, and 42 inches diameter of stones. Particular attention is given to have every mill-stone of even selection of French Buhr, carefully built and balanced, and of the quality of buhr stock adapting it to the work designed, that the best possible result may be attained; as for example, the mill required to grind wheat should be of a different nature from that required for middlings, and different again for corn, &c. The Spindles are made of wrought-iron, having the steel points inserted so as to be easily removed. The bearings of the spindle at the lower end, and around the collar, are of the best and latest improved construction; these with the patent self-tram driving irons and balancing arrangement are of special advantage, and on which the makers have the exclusive right to make. The Husks are of hard wood, and firmly bolted with heavy wrought-iron bolts.

THE COMPLETE GRINDING AND BOLTING FLOUR MILLS so long made by this firm are now in use by more than one thousand millers, although no special effort has been made for their introduction. The smaller custom and merchant mills, when of proper machinery, are proving to be the best investment, even in competition with the large ones. Illustrated Book on Mills sent free.

NORDYKE, MARMON & CO., Richmond, Ind.

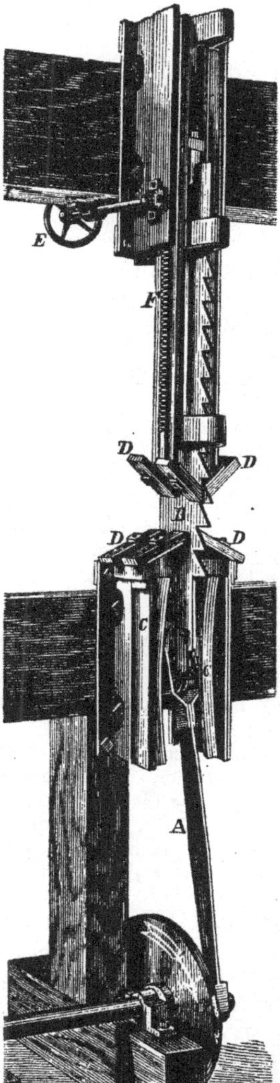

ANDERSON'S

PATENT MULAYS.

Warranted to cut 1-3 more with 1-3 less power than any common Mulay.

Tower's Screw Log Sets

Save the time of one man; allow the saw through without "Stub Short; will set straight or tapers; operated from the head of saw; are the most perfect, accurate and durable in use.

COE & WILKES,

—MANUFACTURERS OF—

Stationary and Portable Steam Engines,

STEAM PUMPS,

MULAY SAW MILLS.

SOLE MANUFACTURERS OF

ANDERSON'S PATENT Oscillating Mulays,

PICKETT'S SAW SHARPENER.

97 and 99 St. Clair St.
PAINESVILLE, O.

SEND FOR CIRCULARS.

OWENS, LANE & DYER
MACHINE COMPANY.

—PROPRIETORS OF—

THE ECLIPSE MACHINE WORKS.

—ARE SOLE MANUFACTURERS OF THE CELEBRATED—

ECLIPSE STEAM SAW MILLS,

—WITH ALL SIZES OF—

Portable and Stationary Engines, Hamilton Steam Thresher, "California Chief."

☞ For any thing Wanted in THE MACHINERY LINE, call and see, or address them, at

HAMILTON, O., *North Hamilton Station.*

ST. LOUIS, MO., *717 North Second Street.*

HOYT'S PATENT MILLSTONE DRESSER.—ON BED-STONE.

Every Miller should know the merits of this Machine. "The greatest invention that has come to the aid of milling for many years.—OTTO TROOST, 13 run mill, *Winona, Minn.*" This machine cuts the furrows out, and faces, with a swiftly revolving emery wheel. It will furrow out a run of stone in one day, and will smooth off the entire face of the stone in less than two hours. It is run by the power of the spindle. For particulars and price list of this important invention, send for circular, free, or 25 cents for *Northwestern Miller* for one year, and circular.

DEAN, SMITH & CO., LA CROSSE, WIS, Sole Proprietors and Manufacturers.

JAMES LEFFEL'S IMPROVED

Double Turbine Water Wheel.

7,000 IN USE,

Under the Highest and Lowest heads utilized by any wheel in this country.

TWENTY-FOUR SIZES MADE,

Comprising the largest and smallest Water Wheels ever applied to manufacturing purposes.

SHAFTING, GEARING, PULLEYS, CORE WHEELS, STONE PINIONS and MILL SPINDLES

Furnished to order.

Illustrated Book, containing full particulars, prices, &c., sent free, postage paid, to parties interested in water power. Address

JAMES LEFFEL & CO., Springfield, Ohio.

Heald & Sisco Patent Centrifugal Pumps

VERTICAL AND HORIZONTAL.

First Premiums at New Orleans, Cincinnati, and New York. "*Medal of Special Award,*" American Institute, 1872.

Perfect satisfaction guaranteed. The cheapest, most durable, popular and successful Pump known, for Paper Makers, Tanners, Contractors, Brick Makers, Distillers, etc. Pumps with engine on frame, complete, at low figures, for Wrecking, Dredging, Irrigating, etc. Illustrated pamphlet, free. 800 references to parties actually using the Pump. 24 pages of the strongest possible testimony. Address

HEALD, SISCO & CO., BALDWINSVILLE, N. Y.

IDA (PAPER) MILLS. OLYMPUS (PAPER) MILLS.

ARKELL & SMITHS,

—MANUFACTURERS OF—

MANILLA PAPER,

—AND MANUFACTURERS OF—

Cotton and Patent Soft-Tie Paper Flour Sacks,

AND JOBBERS OF GRAIN BAGS.

CANAJOHARIE, NEW YORK.

As we manufacture our own paper, and run the mills exclusively on selected Manilla stock, it enables us to produce a uniform

STANDARD MILLING SACK.

We are the sole and only makers of a SOFT TIE Flour Sack, and all sacks of our manufacture are characterized by the CORRUGATED or CRIMPED TOP, enabling the sack to be tied easily, and avoiding the liability of the string cutting the folds produced by the closing of the mouth of the bag.

The bottom of our make of sack, by the evenness and breadth of the laps enables us to secure it in such a manner as to render it impossible to leak.

It is, and will continue to be, our ambition to make our production reliable, knowing that a good article (which is the cheapest) is required by the trade, rather than an inferior one at a lower price.

These goods have been before the public for the past ten years, and we would respectfully refer to the high reputation they have earned as a guarantee of our earnest endeavor to meet the wishes of the trade.

Bookwalter Engine.

CHEAPEST AND BEST ENGINE EVER MADE.

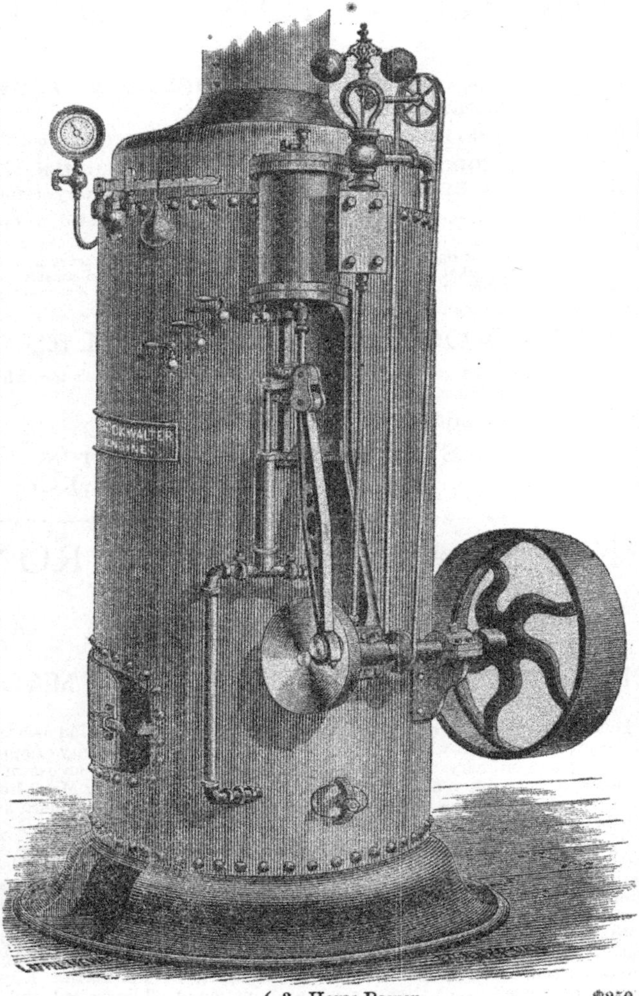

SEND FOR FULL DESCRIPTIVE CIRCULAR.

LOOK AT THESE PRICES. { 3 Horse Power........................ ....$250.00
4½ Horse Power.................................. 300.00

—MANUFACTURED BY—

JAMES LEFFEL & CO., Springfield, Ohio.

James Leffel's Improved Double Turbine Water Wheel.

There is perhaps no surer evidence of practical merit than success long established and widely extended, and based upon repeated trial upon the most exacting conditions. An invention of but little real utility may obtain a temporary reputation by means of shrewd management in bringing it before the public; but its deficiencies will inevitably come to light, and a final verdict will be pronounced upon it in accordance with the facts. Cases in point are of almost daily occurrence, in which a transient popularity is gained by a device which will not endure the test of experience, and which speedily disappears from the market. It is therefore hardly too much to say that the fact that some *seven thousand* of the James Leffel Double Turbine water wheel are now in successful operation, under heads varying from 1½ to 240 feet, and that the demand for them is larger to-day than at any previous period, constitutes the strongest possible evidence that it is what it is claimed to be by its inventor and manufacturers—the most perfect water wheel ever offered for sale.

HOW THE WHEEL HAS BEEN BROUGHT TO PERFECTION.

No machine, however simple, durable and perfect in appearance, will in every respect prove satisfactory, when first put into operation; many parts will require perhaps a change of form, strengthening, or may be an entirely different arrangement, upon application to the work to be performed, and a trial of two or three years. In fact it requires years of diligent study and practical experience,

JAMES LEFFEL'S

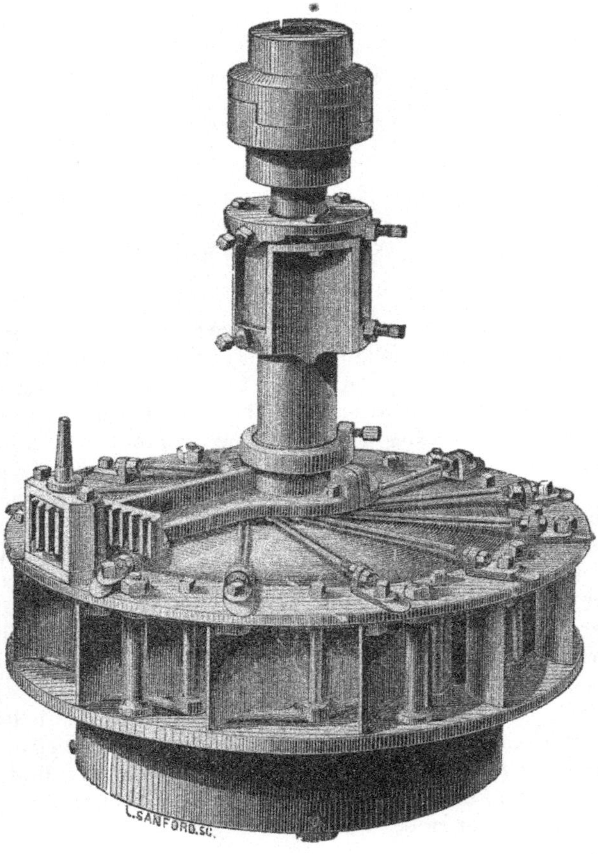

Improved Double Turbine Water Wheel.

particularly with a water wheel, to so perfect all of its parts, as to make it successful under all circumstances, even though it be sound and practical in principle; of course many of them never can, by any amount of labor or attempted improvement, be made to operate all species of machinery, and must always remain but little better than worthless. To the general principle first stated, the Leffel Double Turbine has perhaps been no exception; during its introduction for the first four or five years, many of them were, no doubt, imperfect and not so durably made, as all newly introduced machinery is necessarily. Yet it has been our constant aim and design, to study and know its faults, which so large an experience with it would enable us to discover, and to the proper remedies and improvements we have directed our attention with the most gratifying results. A number of parts have been strengthened, others changed, and some added, all of which we could point out and describe, stating the object, purpose, &c., if we thought it desirable. Among the most noticeable modifications and additions made, (some of which are patented) are the improved link for operating the gates; the process for lining the iron plates with brass or any anti-corrosive metal (applied only when specially ordered); the combination of the toothed segment with the gate-arm in such a manner that the segment can be removed when the teeth become worn, and a new one supplied; the spherical iron penstock; and the improved method of casting solidly in one piece both wheels, by means of which the edge of the diaphragm can be made much thinner and yet stronger, assisting also to separate more perfectly the due proportion of water to each wheel. Half the buckets being made of good boiler iron, and the fillets retaining them being improved both in form and strength, it is impossible to break or tear out any of them; as a result of which, out of the last 4,000 wheels put into operation not one has lost a single bucket. One set of buckets can easily be bolted or riveted to the wheel flange, if it were considered advisable. All such buckets are, however, liable to frequent derangement, by working loose and striking the inside of casing and end of guides often dropping entirely loose and breaking others; subjecting the parties to the expense and inconvenience of taking the wheel from the casing to replace the broken ones. We prefer and recommend only those cast solidly into the wheel, thus enabling them to withstand the shock of blocks, stones and other rubbish to which they are so often subjected, and avoiding also the annoyance of removing the wheel from casing. Practically, the wheel itself is perfect. In fact the durability of the entire wheel and casing is such that the entire amount of repairs called for at all the shops of the firm, per annum, is covered by a sum so extremely small, in view of the fact that about 7,000 wheels are in operation, as to be scarcely worth estimate. The firm have within the last two years so arranged and systematized the process of manufacture that if any part is accidentally broken it can at once be duplicated, another being supplied by express on receipt of the necessary information. In short, the Leffel Improved Double Turbine has kept pace, from its first introduction, with the most advanced developments of mechanical science; and for any purpose for which the power of water is employed, it may be safely guaranteed as having no equal in utility, economy and durability.

DOUBLE WHEELS.

An idea exists to a considerable extent, that water wheels may be so constructed, with two or more sets of buckets, in such a manner that each set of buckets may form a separate wheel, and that the water may be received first by one set of buckets, or one wheel, and after passing from the first, then to operate on a second arrangement of buckets, or wheel, and so on with as many sets or wheels as there may be, or until the last one is passed or operated upon; thus in their opinion, obtaining a much greater percentage of the power of water, than is ordinarily utilized by the use of well constructed wheels of other kinds. In fact a much greater power is often claimed for them than can possibly exist in the quantity of water used. Again, there is another class of wheels claiming to be double wheels, which are in reality and principle, but single wheels; their builders believing by such representation that the reputation and popularity of our wheel, (so celebrated for its truly double character) mây thus directly benefit them. A single wheel, either a central or vertical discharge wheel, is commonly used, with a partition through the middle of the tier of buckets, thus only dividing the wheel, without in the least changing the action of the water on the buckets on either side of the partition or division, and without any modification of the principle of construction.

The Leffel Double Turbine should not be confounded with either of these classes of wheels, as it is constructed, and acts, upon entirely and essentially different principles, which are peculiarly characteristic of it as a water wheel. There is in it a combination of two independent sets and kinds of buckets, one a vertical, the other a central discharge, each entirely different in its principle of action upon the water, yet each wheel or series of buckets receiving its water from the same set of guides at the same time; but the water is acted upon but once, since half of the water admitted by the guides passes to one wheel, and the other half of the water to the other wheel; the water leaving both wheels or sets of buckets at the same time and as quickly as possible. These two sets of buckets are so combined as to make really but one wheel; that is both are cast in one piece and placed upon the same shaft. By this arrangement there is admitted the greatest possible volume of water, consistent with its economical use, to a wheel of any given size, and at the same time the greatest area for the escape of water is secured. The surface in the wheel is thus reduced to minimum as compared with the quantity of water used, avoiding a very material loss by friction, which otherwise seriously diminishes the working power of a wheel. The value of this arrangement will be fully appreciated by those who who understand the practical effect of the frictional surface in a water wheel.

JAMES LEFFEL & CO.

SPRINGFIELD, OHIO.

Leffel's Improved Patent Globe Casing.

The plate on page 335 represents our New and Improved Patent Globe cast iron Penstock, or Casing, which we are making, and in which many of our wheels are now placed. The form being that of a Globe or Sphere, it at once secures the greatest strength, with the least weight, and at the same time affords the largest space for the water to circulate above and around the wheel; while it also admits of the smallest exterior dimensions, and therefore occupies less space, than any other form or shape that can be adopted. As none of the parts are subject to wear or breakage, it never requires replacing, and of course its durability is beyond question.

It is, as the plate represents, two hemispheres bolted together, thus enabling it to be easily taken apart, if at any time it should become necessary. There is a movable cap or cover, C, bolted on the top of the casing, which can at any time be removed, (when the head of water is not standing in the case) and the wheel lifted bodily out of the casing, the opening in the top of same being amply large for that purpose, though it is seldom necessary to remove the wheel from any cause. There are two large man-holes on the sides, also a hand-hole B on the top cover, through which any obstruction can be removed, that may by carelessness or accident get into the casing; through these holes the wheel can also at any time be examined. On the top of cap C is bolted firmly a bridge-tree, carrying a good, broad oil bearing, for the support of the upper end of the water wheel shaft, to which a clutch coupling D is attached, immediately above said bridge-tree. In the cover or cap C are arranged neat, snug and tight stuffing boxes through which the gate rod, A, and water wheel shafts pass, and by which any water is prevented from discharging; they are supplied with tightening bolts by which they

LEFFEL'S IMPROVED

PATENT GLOBE CASING.

can be tightened down, should the packing at any time become worn or loose. In fact, the whole affair, when well set and arranged, is perfectly water tight, not leaking a drop, and could be located upon a floor near to any of the machinery if desired.

They cannot be frozen up, since the iron is thick, and the circulation of water always sufficient to prevent freezing. A short tube or cylinder is attached to the bottom, which is intended to be slightly submerged under standing tail water; it has a flange with its face turned and with bolt holes, to which an iron draft tube can be attached, and by a little care a perfectly air-tight joint can be made; in cases where such draft tube is used of course the entire casing can be set higher, and sometimes in a more convenient location.

We cannot say too much in praise of this casing, particularly for high falls; being made strong and water tight it will always remain so. It has been fully tried and tested under almost every circumstance, and has proven in the highest degree satisfactory; some of them are under heads from 80 to 240 feet, and stand the tremendous pressure admirably. In fact almost all of our small wheels up to 20 inches diameter are now ordered by our customers to be encased in this manner, such has been the satisfaction they have given. Of course it is not absolutely necessary to use it, except in particular instances, where a want of space or other circumstances would prevent the erection or use of a wood flume or box in which to place the wheel; but at any time and under almost any condition it is preferable and makes a number one arrangement, especially in any case whatsoever where the power is taken off below the level of head water. But its greatest convenience is in locating wheels under mills, and in other difficult places, where posts, foundations, walls, &c., cannot be removed; such difficulties being obviated by the compactness of its form, and the ease with which it can be connected to the head water by a pipe of suitable size.

JAMES LEFFEL & CO.

SPRINGFIELD, OHIO.

Spotted something that isn't quite right?

Let us know and we'll fix it for future reprints. And as a thank you for your diligence, we'll send you a free Sothis book of your choice.

publisher@sothis.press

This page intentionally left blank

Made in the USA
Monee, IL
08 November 2026

67812143R00204